The Mechan
Design Process
Case Studies

David G. Ullman
Professor Emeritus, Oregon State University

THE MECHANICAL DESIGN PROCESS CASE STUDIES

Published by David G. Ullman, 621 Aeronca St., Independence Oregon 97351. Copyright 2017 by David G. Ullman. All rights reserved. Printed in the United States of America. No part of this publication may be reproduced or distributed in any form or by any means, or stored in a database or retrieval system, without the prior written consent of David G. Ullman, including, but not limited to, in any network or other electronic storage or transmission, or broadcast for distance learning.
ISBN 978-0-9993578-1-1

Ullman, David G., 1944- author.
The mechanical design process / David G. Ullman, Professor Emeritus,
Oregon State University.
ISBN 978-0-9993578-1-1   1. Title.
1. Machine design.
  TJ230.U55 2017

www.mechdesignprocess.com

ABOUT THE AUTHOR

David G. Ullman is an active product designer who has taught, researched, and written about design for over thirty years. He is Emeritus Professor of Mechanical Design at Oregon State University. He has professionally designed fluid/thermal, control, and transportation systems. He has published over twenty papers focused on understanding the mechanical product design process and the development of tools to support it. He is founder of the American Society Mechanical Engineers (ASME)—Design Theory and Methodology Committee and is a Life Fellow in the ASME. He holds a Ph.D. in Mechanical Engineering from the Ohio State University.

Contents

Preface

This book contains a series of case studies that demonstrate how industry makes use of the best practices found in the text *The Mechanical Design Process.* Case studies have long been used in business schools but only rarely in engineering studies. However, studying how professionals have solved design problems is one window on the design process that is useful to students and practitioners.

What is unique about these case studies is they are not purely academic, but are co-written by practicing engineers to show real world examples of good practice.

The case studies range from those featuring very small companies to NASA and BMW. They range from the design of bicycles to heat exchangers. They include systems that solely of hardware to those with integrated electronics and software.

The process of developing the case studies varies, but generally follows the same pattern. I interview interested engineers involved in the development of a current or past product. During the interview we identify the best practices that were used either formally or informally during the development of the system. From this, I write a draft of the case study and iterate with the engineers as they add photos or other graphics to support the text. When it is complete we get the needed management approvals. It is then available for company use and for inclusion here.

The case studies are also are individually as a pdf download for a fee at the book's website, www.mechdesignprocess.com.

If you are a practicing engineer, and you or one of your colleagues are interested in working with me on a new case study, please contact me at ullman@davidullman.com or through the book's web site.

David G. Ullman
2017

From Constraints to Components at Marin Bicycles
A Case Study for The Mechanical Design Process

Introduction

This case study details the development of the Marin Mount Vision Pro mountain bicycle rear suspension. Marin Bicycles is one of the earliest developers of the mountain bicycle as we know it today. Founded in 1986 by Bob Buckley (who still is active in the company) they are still leaders in mountain bicycle innovation.

The Mount Vision Pro was developed in 2006- 2007 by a team at Marin led by Jason Faircloth, a young mechanical design engineer. It was introduced in 2008 to good reviews and has sold well. The mountain bicycle market is highly competitive with industry leaders such as Marin pressured to develop new products each year in time for the annual bicycle shows. Within a year or two other manufacturers will copy and adopt any new technologies developed by companies like Marin.

Figure 1: The Marin Mount Vision Pro

Additionally, Marin had been working for a number of years to make a breakthrough in rear suspension design as will be described. Thus, the development of the Vision Pro suspension was a combination of technology push and market pull for new products with better performance.

The Marin Mount Vision Pro bike was designed for the Cross Country mountain bike enthusiast. It is a quality and fairly expensive bicycle (over $3,000USD). The primary demographic for this bicycle is male, 25-50 years old. But, because of its modern look and marketing it is also designed attract female and other age groups riders. It is intended for use on technical trails where there is a mix of uphill and downhill, where light weight and pedaling efficiency are of primary importance.

The Problem: Marin needed to design the rear suspension for their new Mount Vision Pro bicycle. This was a more complex suspension than they had designed before.

The Method: Marin used a structured method that progressed from Constraints to Configurations to Connections to Components. This methodology helped them ensure that the final configuration met the needs. Each of the four steps is described here.

Advantages: This method forces rigor and eliminates surprises. There is little down-side other than taking longer up front.

Constraints, *Configurations, Connections, Components*

The first step in this method is to understand the spatial constraints for the system. For the rear suspension of a mountain bicycle, the spatial constraints are shown in Fig. 2. Beyond the obvious need to connect the wheel to the frame, the Marin engineers also wanted to control the path the wheel made relative to the frame as the suspension deflected, the stiffness of the suspension, and the chain length.

Figure 2: Physical constraints for the Mount Vision

Ideally, the wheel of the bicycle should move "nearly" straight up and down as it deflects. If the suspension was designed as a simple bar with a single pivot as on many bikes (see fig 3), then the wheel would make an arc with it moving closer to the front of the bike as it deflected. This would give the rider the feeling she was falling backward as the wheel moved upward. The Marin engineers wanted to control the wheel path to manage the feel transmitted to the rider. As important as the wheel path, was the change in stiffness. The ideal suspension system for any vehicle is soft, i.e. has low stiffness, when it goes over small bumps and gets stiffer for large bumps. In other words, the larger the deflection, the stiffer the suspension system should become.

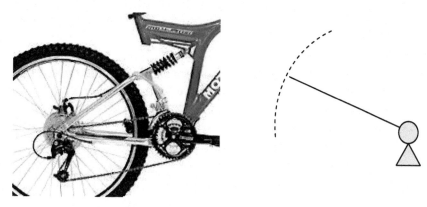

Fig 3: A simple, single pivot suspension

Finally, there was the desire to control the chain length. Consider a suspension that was designed so that when the pedals were pressed, the resulting tension in the chain pulled the suspension up (i.e. the frame down). The rider, when feeling the frame drop would then ease off the force and subsequently the frame would rise. Feeling the frame rise, the rider then reapplies the pedal force resulting in a "pogo" motion and a very uncomfortable ride. Pogoing is often seen on poorly designed suspensions. Thus, an additional constraint is that the motions and accelerations felt by the rider will not lead to poor suspension performance.

Summarizing, the spatial constraints for the rear suspension are:
- Wheel and chain must clear frame for all deflections
- Wheel should move in a designed path
- Low stiffness for small deflections, increasing with deflection
- Chain length should not change during deflection

Constraints, **Configurations**, Connections, Components

The second step is to develop the configuration or architecture of a candidate system. The simplest type of suspension that can be put on a bicycle is a one with a single pivot as shown in Fig 3. On that bike, the pivot is near the center of the crank and every point on the rear triangular structure (called the rear "stay") rotates around this point. As the wheel deflects, it makes a circular arc and the chain gets shorter, violating two of the spatial constraints. As the wheel moves up, the shock gets shorter. Shocks on bicycles generally have an air or oil damper with a mechanical, coil spring wrapped around it. This spring has a stiffness that remains essentially constant as the wheel deflects. Thus, it is clear that this type of suspension will not work for the Marin Mountain Vision Pro.

The technical advancement developed by Marin was to use a 4-bar linkage called the "Quadlink". The Quadlink was not the first 4-bar suspension used on a mountain bicycle, butt it did bring this type of mechanism to a high level of refinement.

On the Quadlink, the rear stay, the connection point for the rear wheel, rotates about the instant center and makes a nearly straight line. In order design the shape of the path followed the wheel Jason specified the lengths of links and the relative positions for the two fixed pivot points (the distance between them and angle of the line connecting them), for a total of 7 variables. There was a lot of design freedom.

To design the Quadlink, Jason used the Autodesk Inventor Dynamic Simulation capability combined with an Excel Model that helped in parametrically studying the seven variables that determine the linkage to meet the spatial constraints. The final design is shown in Fig. 4. Inventor helped Jason model and see the motion as the suspension deflected.

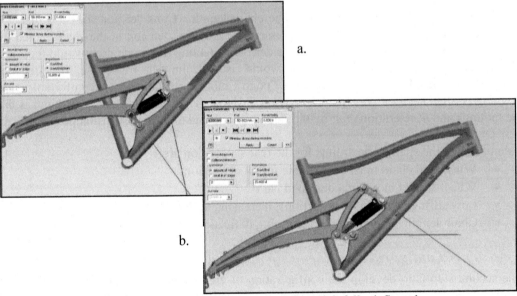

Figure 4: Simulation of the Quad link suspension, a) undeflected, b fully deflected.

The resulting linkage gives a fairly straight wheel path with near constant chain length. Further, by controlling the location of the virtual center (where the black lines cross) and the positioning of the shock he was able to achieve low stiffness for small deflections, increasing with deflection. Specifically, when the virtual center is nearly under the crank (4a) the moment arm of the rear stay is much shorter than when the suspension is deflected (4b).

Constraints, Configurations, **Connections**, Components

The third step is to design the connections. On the Marin Mount Vision Pro, the connections are those between the links in the 4-bar linkage, those connecting the shock to the bike and those that connect the fixed parts together. Considering Fig 4, the shock could have been mounted in many different ways- between any two elements that move closer together as the system deflects. The addition of the shock adds two more pivots to the assembly making a total of 6 pivoting connections.

The Marin engineers reduced the number of pivots by mounting the shock on existing linkage pivots as shown in Fig. 4. As the suspension system deflects, the two pivots move toward each other. In fact, Jason, when determining the lengths of all the seven members, took the needed change in length of the shock as an additional constraint. The decision to mount the shock in this manner made the design of linkage more challenging and connections more complex, but the tradeoff for fewer pivots made this worthwhile.

The two shock pivots need to have the link and shock free to rotate about the axel (shown as a centerline in Fig. 5). Note in Fig. 4, the amount of rotation of these elements is small, only a few degrees in some cases. Bearings that operate primarily in one position and only move a small amount from that position present their own design problems as small deflections do not force the lubricant to flow to all the areas.

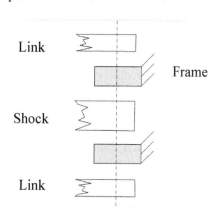

Figure 5: The components in shock pivots

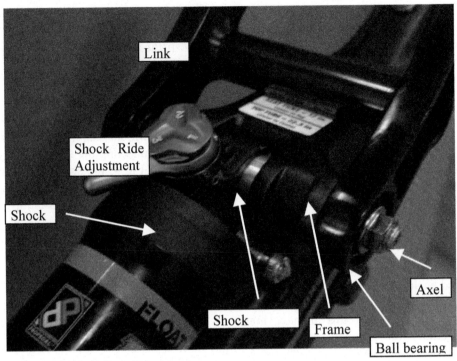

Figure 6: Final design of Pivot 2

The final connection at one pivot is shown in Fig 6. Connections between components that are moving relative to each other need to be addressed, as the components are refined in the next section.

Constraints, Configurations, Connections, **Components**

Finally, the actual components were developed. Vision Pro parts needed to be light in weight, manufacturable in volumes that matched the sales projections and had a look that would attract sales. Thus, these parts were a combination of structure and eye candy. .

The link (identified in Fig. 6) is a very simple component that, like many on the bike, is forged aluminum with the bearing mounting surfaces machined. It is shown in two views in Fig. 7.

The bearing between the axel and the link, shown pressed into the link in Fig. 7, is a rolling element ball bearing. As mentioned earlier, this bearing does not rotate very much and thus requires special consideration. The final bearing chosen was one that was specially designed for aircraft control systems, another application with small, repetitive motions.

Figure 7: Link A

The link was modeled with Autodesk Inventor's FEA capability, so that the stresses in it could be seen during dynamic simulation. Since the loading signature on a mountain bike is not well modeled, FEA primarily served as visualization and learning tool with final decisions made based on physical fatigue testing.

The rear stay components could have been made out of round aluminum tubes welded together as with most aluminum bikes. However, to get a better "look", the designers wanted tubes that curved, and to save weight, the engineers wanted tubes that tapered. As shown in Figs. 1 and 4 these two requirements were met. The manufacturing method used is called hydroforming. To hydroform, a round tube is put in a die and then the tube is filled with high-pressure liquid causing it to deform and be shaped by the die.

Summary

During the design of the rear stay a conscious effort was made to consider the Constraints, Configurations, Connections, and Components in a reasonable linear manner. This helped ensure that some details were not addressed too early in the process and that there was a clear path to follow.

Acknowledgements

Jason Faircloth of Marin Bicycles assisted in the development of this case study

Autodesk® Inventor® sponsored the development of this case study.

Multi-duty PC Boards at Sound Devices
A Case Study for *The Mechanical Design Process*

Introduction

Sound Devices designs and manufactures equipment that records the audio you hear in movies. The 788T High Resolution Digital Audio Recorder is a powerful eight input, twelve-track digital audio recorder designed for production sound. It records and plays back audio to and from its internal drive, CompactFlash, or external drives, making audio recording simple and fast. This unit, introduced to the market in April 2008, is half the size and weight of the competitor's products of similar capabilities. It is so compact that can easily be carried by sound professionals in an over-the-shoulder bag.

A unique feature of the 788T is that printed circuit boards (PC boards) not only serve their intended purpose for mounting and connecting electronic components but also the physical mounting and support for the controls. This dual use of the PC boards has resulted in a very efficient design from an assembly viewpoint.

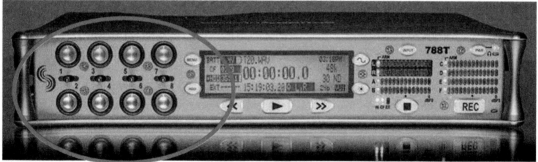

Figure 1: Sound Devices 788T

This case study will focus on the design of the Light Ring Assembly, the set of controls circled in Fig. 1. Specifically:

The Problem: Design the Light Ring Assembly, an assembly in the 788T High Resolution Digital Audio Recorder, which is mechanically and electronically easy to assemble.

The Method: The Light Ring Assembly was designed on Autodesk® Inventor® taking into account best Design for Assembly (DFA) practice.

Advantages: Even though this is a low volume product, the application of DFA methods resulted in a light, efficient design.

The 788T

On a movie set, the soundman's job is to record audio for mixing and editing later. Using the 788T she may record up to 8 separate channels of audio. Often the 788T is carried in a bag as shown in Fig. 2 so it must be light, small and durable as it gets banged around a lot. 788Ts have been used in rainforests, on top of Mt. Everest and in every other conceivable environment.

Figure 2: Soundman with 788T in his bag

Once the 788T is connected to microphones the soundman uses the controls to set the levels for recording. As shown in Fig. 3 (from the user's manual), the channels are activated and gains are set by using the eight pop-out potentiometers (called the Input Gain Controls) on the front of the unit.

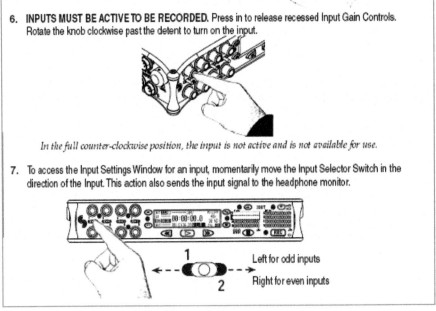

6. **INPUTS MUST BE ACTIVE TO BE RECORDED.** Press in to release recessed Input Gain Controls. Rotate the knob clockwise past the detent to turn on the input.

In the full counter-clockwise position, the input is not active and is not available for use.

7. To access the Input Settings Window for an input, momentarily move the Input Selector Switch in the direction of the Input. This action also sends the input signal to the headphone monitor.

1   Left for odd inputs

2   Right for even inputs

Figure 3: Details from the 788T Users Manual

Using the 3-position Input Selector Switches, the soundman can listen to each of channels, one-at-a-time to hear that he has the level set so that a good signal is recorded. The Input Selector Switch is normally in the middle position, but by moving it to the left or the right, the user can select either of the two channels to send the input signal to the headphone monitor.

The Light Ring Assembly

On the 788T, the Input Gain Controls and the Input Selector Switches are part of the Light Ring Assembly. This assembly (shown in Figs. 4 and 5) was designed by Jason McDonald, a young mechanical engineer who joined the Sound Devices staff in 2007. His addition has allowed them to produce equipment that is smaller and lighter.

The Input Gain Control (IGC) potentiometers for inputs 1-4 are mounted on the Upper Horizontal PC Board and those for inputs 5-8 on the Lower Horizontal PC board as can be seen in the exploded view in Fig. 5. All the potentiometers are held to the board by soldered tabs and electrical connectors. Circuitry on these boards supports the potentiometers. Also on each of these boards is a connector that allows the horizontal boards to electrically plug into the Vertical PC Board.

The vertical PC board is populated with the four 3-position Input Selector Switches, a set of LEDs for each input

Figure 4: X-ray Isometric of the Light Ring Assembly

channel and driver chips for the LEDs. The LEDs surrounding the Input Gain Controls indicate the input activity for each channel. There are nine LEDs distributed in a circle with three red LEDs indicating that the signal is being clipped, three yellow LEDs to indicate the peak setting and three green LEDs to indicate that the channel is in use and the signal is within set limits. The LEDs feed their light into the Light Ring as will be described in a moment.

The Horizontal Boards are mechanically fastened to the Vertical Board using the Light Rings. These are injection molded polycarbonate parts that serve multiple purposes.

The section view of the Light Ring (Fig. 6) shows its important features. The LEDs on the Vertical PC Board fit into the groove in the back of the Light Rings where the matte finish disperses the light in the Ring. They then transmit the LED light to the front of the unit and mix the light from the 3 same-colored LEDs to make an even circle of red, yellow or green light.

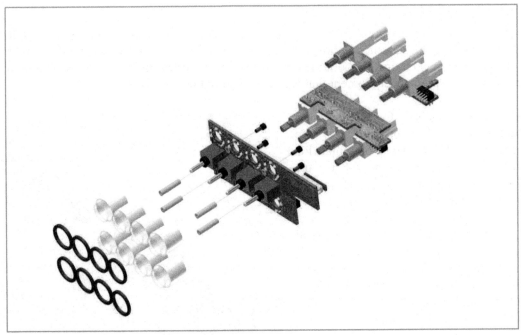

Figure 5: Light Ring Assembly Exploded View

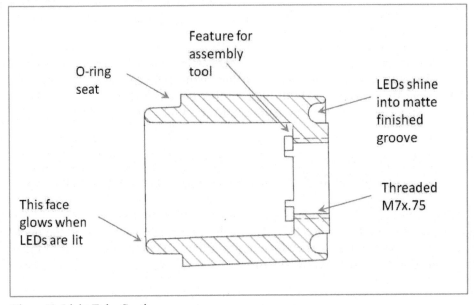

Figure 6: Light Tube Section

Each Light Ring threads onto a potentiometer and holds the horizontal PC boards to the vertical PC board. The Light Rings also have a groove in them to hold the O-ring (shown in Figs. 5and 7) used to keep water and dirt out of the unit. Finally, there are features molded into the Light Ring making the assembly easy. Jason also designed a special driver that mates to these features, shown in Fig. 7.

The entire Light Ring Assembly itself is mechanically mounted to the faceplate of the Recorder with four standoffs that are screwed to the vertical PC board. These can be seen in Fig. 5.

Figure 7: Light Tube with Driver

Design for Assembly (DFA) Evaluation

The cleverness of this design can be fully appreciated by performing a Design for Assembly evaluation on it. Jason did not use a formal DFA process during the design of the assembly, so a formal evaluation is worthwhile. The evaluation used here is developed in Chapter 11 of The Mechanical Design Process. A template for this evaluation can be downloaded for free from the book's web site. The template is a Word document as shown in Fig. 8. The scores shown in the diagram were entered by the author of this case study after dissecting the Light Ring Assembly. As described in The Mechanical Design Process, the scoring is useful to guide designing. However, the value of the score itself is only useful in comparison with other, alternative configurations. To make this more interesting, parallel to describing the logic for the 788T Light Ring Assembly scores, an alternative, more traditional mounting scheme will be scored for comparison. A full description of each of these measures can be found in the reference.

Generally DFA becomes important for high volume products where the assembly costs may be a significant part of the entire cost. While the 788T is not high volume, the DFA analysis shows the quality of the design.

The evaluation for the Light Ring Assembly assumes that the electronic components are already mounted on the PC boards and so the Bill of Material (BOM) for the final assembly is:

Part	Req.
Upper horizontal PC board	1
Lower horizontal PC board	1
Vertical PC board	1
Light Rings	8
O-rings	8
Stand-offs	4
Screws for stand-offs	4
Foam rubber pads	4

These parts are shown in Fig. 5, with the exception of the foam rubber pads that go on the Input Selector Switches to keep out water and dirt as do the O-rings at the Light Rings. Also, not counted or shown are the screws to attach the assembly to the faceplate.

On the form in Fig. 8 the wording for each of the 13 measures are different than in the text where they are given as design guidelines. The guidelines version of the wording is used below as each of the measures is applied to the Light Ring Assembly.

Overall Assembly

Guideline 1: Overall Component Count Should Be Minimized. The first measure of assembly efficiency is based on the number of components or subassemblies used. This is accomplished by comparing the minimum number of components possible to the actual number used. For the Light Ring Assembly, the total number shown in the BOM is 31 parts. In deciding the minimum number of parts necessary for the assembly, consider that the material properties of the PC boards must be significantly different from the Light Rings and that the O-rings and foam pads yet different again. Further, even a single PC board must fasten to the chassis in at least 3 places. Then, the minimum number of parts is 8 Light Rings, plus 8 O-rings, plus a single foam pad, all the electronics on a single assembly and 3 snap parts to hold the PC board, totaling 21 parts. This gives an improvement potential of 32% which is considered good as indicated on the DFA work sheet, giving 4 points.

(DFA) Design for Assembly

Assembly Evaluation for: 788T Light Ring Module	Organization Name : Sound Devices

ERALL ASSEMBLY	
rall part count minimized	Good
imum use of separate fasteners	Very good
e part with fixturing features (Locating surfaces and holes)	Very good
ositioning required during assembly sequence	>=2 Position
embly sequence efficiency	Very good

RT RETRIEVAL	
racteristics that complicate handling (tangling, nesting, flexibility) have been avoided	Most parts
s have been designer for a specific feed approach (bulk, strip, magazine)	Some parts

RT HANDLING	
s with end-to-end symmetry	All parts
s with symmetry about the axis of insertion	All parts
re symmetry is not possible, parts are clearly asymmetric	All parts

T MATING	
ght line motions of assembly	All parts
mfers and features that facilitate insertion and self-alignment	Some parts
imum part accessibility	Most parts

ly for comparison of alternate designs of same assembly	**TOTAL SCORE**	78

ber: Jason	Team member:	Prepared by: David Ullman	Date: Sept 1 2009
ber:	Team member:	Checked by:	Approved by:

cal Design Process
08, McGraw Hill

Designed by Professor David G. Ul
Form # 21.0

ign For Assembly (DFA) Analysis for Light Tube Assembly

Guideline 2: Make Minimum Use of Separate Fasteners. One way to reduce the component count is to minimize the use of separate fasteners. This is advisable to ease assembly, save cost and reduce stress concentrations. Eliminating fasteners is more easily done on high-volume products, for which components can be designed to snap together, than on low-volume products. Here Jason has used the Light Rings to both channel the LED light to the front panel, to spread the light around the whole circumference, and to mechanically hold the PC boards together. The 8 Light Rings are the only mechanical connection in the assembly.

Separate stand offs are used to hold the assembly to the chassis. This is seen as an opportunity for further improvements. Based on the cleverness of the Light Ring design, the use of separate fasteners measure is very good, resulting in 6 points on the DFA worksheet.

Guideline 3: Design the Product with a Base Component for Locating Other Components. This guideline encourages the use of a single base on which all the other components are assembled. The Vertical PC Board serves as this base and all the other parts fit into through holes, or, electrically, with connectors. Thus, this measure is very good, adding 6 more points to the worksheet.

Guideline 4: Do Not Require the Base to Be Repositioned During Assembly. This measure is important if automatic assembly equipment such as robots or specially designed component placement machines are used during assembly. Here the volume is low enough that hand assembly is used, but still, this measure indicates how much the base part (the vertical PC board) needs to be manipulated during assembly. The assembly sequence begins with the vertical PC board face down where the screws that hold on the stand offs the horizontal PC boards are inserted. Then the whole assembly must be rotated 180deg to insert and screw in the Light Rings and insert the O-rings and foam rubber.

Using the scale on the DFA worksheet, the two positions add 4 more points.

Guideline 5: Make the Assembly Sequence Efficient. If there are N components to be assembled, there are potentially N ! (N factorial) different possible sequences to assemble them. For this assembly, the sequence described in the repositioning analysis seems very good, for 6 points.

Evaluation of Component Retrieval

The measures for retrieving components range from "all parts" to "no parts" on the DFA worksheet. If all components achieve the guideline, the quality of the design is high as far as component retrieval is concerned. Those components that do not achieve the guidelines should be reconsidered.

Guideline 6: Avoid Component Characteristics That Complicate Retrieval. Three component characteristics make retrieval difficult: tangling, nesting, and flexibility. Only the wire lead off the back of the Vertical PC Board fits this description and so "most parts" is indicated on the DFA Worksheet, giving 6 points.

Guideline 7: Design Components for a Specific Type of Retrieval, Handling, and Mating. The 788T is manually assembled and so parts should be designed with this type of assembly in mind. Although there are no special features on the parts to accommodate human hands, the parts are relatively easy to handle and so a neutral score of "some parts" is given, 4 points.

Evaluation of Component Handling

The next three design-for-assembly guidelines are all oriented toward the handling of individual components. They will be treated together.

Guideline 8: Design All Components for End-to-End Symmetry.
Guideline 9: Design All Components for Symmetry About Their Axes of Insertion.
Guideline 10: Design Components That Are Not Symmetric About Their Axes of Insertion to Be Clearly Asymmetric.

The end-to-end symmetric parts are the stand-offs, O-rings, and foam rubber pads. There is no end-to-end differentiation for these parts. The bolts that fasten on the stand-offs are axis-symmetric and they are symmetric about their axis of insertion... All other parts are asymmetric and can only go together one way. Thus, all three measures are as good as can be and so they are given an "all parts" score of 8 points.

Evaluation of Component Mating

Finally, the quality of component mating should be evaluated.

Guideline 11: Design Components to Mate Through Straight-Line Assembly All from the Same Direction. This guideline, intended to minimize the motions of assembly, has two aspects: the components should mate through straight line motion, and this motion should always be in the same direction. For the Light Ring Assembly "all parts" meets this, giving 8 points.

Guideline 12: Make Use of Chamfers, Leads, and Compliance to Facilitate Insertion and Alignment. To make the actual insertion or mating of a component as easy as possible, each component should guide itself into place. The potentiometer projections guide the horizontal boards into the vertical board resulting in a "some parts" score, leading to 4 points.

Guideline 13: Maximize Component Accessibility. This guideline is oriented toward sufficient accessibility to allow for grasping and absence of needing to insert parts in an awkward spot. Most parts in the assembly have good accessibility, for 6 more points.

The total score for this assembly is 78 out of a possible 100. This is very good for a product that was designed for low volume. But, as said at the beginning, this number has meaning only when compared to another potential assembly.

Summary

The 788T has been a very successful product for Sound Devices. The quality of the mechanical design has been explored by evaluating one assembly using Design for Assembly (DFA) metrics. Although the designer did not use this exact method in his design effort, it is clear that he knew the DFA guidelines and used them well.

Acknowledgements

Jason McDonald of Sound Devices LLC., Reedsburg Wisconsin assisted in writing this case study.

Autodesk® Inventor® sponsored the development of this case study.

Spiral Product Development at Syncromatics

A Case Study for *The Mechanical Design Process*

Introduction

This case study focuses on the development of Solar Powered Shelter Signs by Syncromatics Corp. Syncromatics is a provider of Intelligent Transit Systems (ITS), specializing in bus tracking, automated passenger counting, passenger information systems and route analytics. Solar powered shelter signs are erected to communicate bus arrival times and destinations to riders waiting at a bus stop. With these signs, riders have clear information about what to expect and when.

Syncromatics solar-powered shelter signs operate over cellular networks and require no wiring, making installation a minimal effort. Further, the signs accommodate both visually and hearing impaired passengers, and are built to the specifications defined in the Americans with Disabilities Act (ADA).

One aspect of this shelter sign is that it is composed of mechanical, electronic and software components, each doing part of the sign's functions. The design process used for each of these fields is somewhat different.

In this case study:

The Problem: Design a product composed of mechanical, electronic and software systems in a coordinated and timely manner.

The Method: The design process is a mixture of methods where hardware and electric circuitry are fixed early and are robust enough to allow agile development of the software.

Advantages: This method commits in sequenced fashion, mechanical, then electronic and finally software elements. This is especially advantageous as variation needs can be met by software changes that can easily and rapidly be made.

Figure 1: The Solar Powered Shelter Sign

The Solar Powered Shelter Sign

The solar powered shelter sign functions as shown in Fig. 2. It collects power from the sun and stores it in a battery inside the housing. The battery is large enough and the power demands are small enough that the sign can function for 6-7 sunless days. The battery powers a cellular receiver which collects information from a system that tracks bus locations and identification. Based on this signal, the time until the bus arrives can be computed by the internal computer (IC Board) and the information displayed for the riders to see. Typical messages are of the form, "Bus 12 to Mango Heights will arrive in 4 minutes". The display is ADA (American with Disabilities Act) compliant with 3 inch characters (one line) or 2 inch characters (for two lines). For visually impaired riders there is a large button on the pole, which when pushed reads the message using an annunciator (a system that translates the text signal to spoken words).

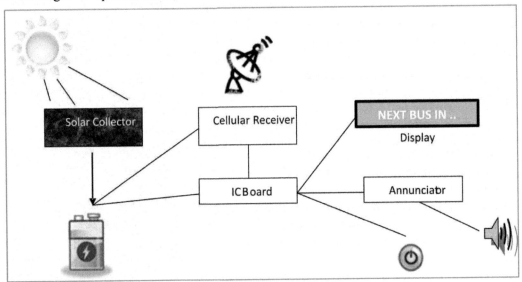

Figure 2: The system

The heart of a system like this is the software that converts a cellular signal to one that can be read and heard. The software, the display and the sound systems are all on battery powered printed circuit boards. All of the components described are connected by the physical structure. This structure must:

1. Physically support the display, PC boards, battery solar cells and cellular receiver,
2. Protect all the electronics from the weather,
3. Provide channels to connect the solar panel and the button to actuate the annunciator to the other components in the main housing,
4. Be tamper proof,
5. Be manufacturable in small quantities.

Since this is a maturing product, and one that may need to be customized for different customers, there will be changes late in the design process and possibly once the product is in service. Thus, a further goal is to design the product so that changes can easily be made.

The Development Process

The process to develop the software, electronics and mechanical components can best be described as three separate but interdependent spirals. Spiral processes have become very popular in software design and, as shown in this case study, can equally be applied to electronic and mechanical systems. Basically, spiral development begins at the center with initial requirements, shown in Fig. 3. From these, some basic concepts are developed and tested with the first prototypes. These prototypes are evaluated for function and the risks identified (i.e. those problems that need to be resolved during product development). The results of this evaluation focus on the next steps, and the next loop of the spiral begins.

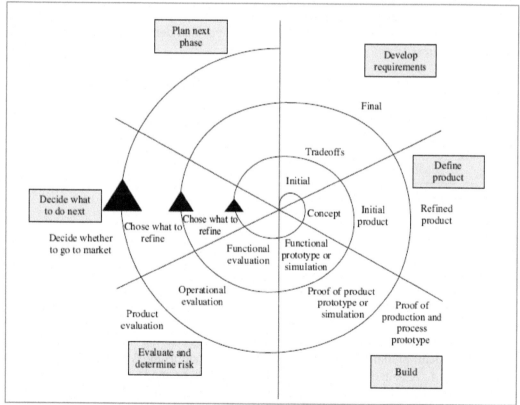

Figure 3: The spiral process (Figure 5.15 in "The Mechanical Design Process" 6th edtion)

21

In ideal practice the initial "Develop Requirements" phase would result in a complete set of specifications for the product. However, for immature products such as the shelter sign, this is not possible and the requirements co-evolve. Syncromatics did a careful job of the initial definition and only made changes identified in previous spirals. Concepts were developed, built and tested from these.

One important aspect of the spiral methodology is that each generation of prototypes moves the system closer to production. Early prototypes need to allow for quick and easy changes. It is the speed with which prototypes can be developed, tested and changed that determines how fast the product can be developed. Also, each loop is based on identified problems and refined requirements. This gives flexibility while keeping the process under control.

In the diagram, the "build" phase is characterized by the type of prototypes developed. Detailing these types can help explain the differences in the three spirals (Table 1). A key factor in this table is how long it takes to make changes. It is the changes that define each loop of the spiral and the time needed for these changes that determine how fast the design process can progress.

	Prototypes			
	Functional	Product	Process	Production
Mechanical Spiral	CAD Drawings Changes in hours	Custom hand built Changes in weeks	Final production Changes in weeks	No changes
Electronic Spiral	Breadboard Changes in minutes	Custom PC board Changes in days	Final PC board Changes in days	No changes
Software Spiral	Code on computer Changes in minutes	Code on board Changes in minutes	Final code Changes in minutes	Change to meet customer needs

Table 1: Prototypes used in the development of the Solar Powered Sign

In the first spiral the goal was to show that the sign worked. Thus, proof-of-concept or proof-of-function prototypes were developed to demonstrate the function of the product in comparison with the customers' requirements and engineering specifications. For the mechanical system, preliminary CAD drawings were made using Autodesk Inventor® , as shown in Fig. 4. Changes to these solid models could be made in hours, aiding in the development to meet the structural requirements itemized earlier. Note that on previous projects, prior to using solid modeling tools, functional prototypes were made with cardboard or thin metal, and any changes took much longer.

Functional prototypes for the electronics were made on breadboards (plastic boards designed to easily mount components and run wires between them) or mocked up with off the shelf components (e.g. the cellular antenna and the display). Using a breadboard, changes could be made in minutes making concept development modifications easy to explore. Finally, software was coded into a desktop computer so that its function could be simulated and, as with most code, changes could be made in minutes.

On the second spiral, proof-of-product prototypes were developed to help refine the components and assemblies for the mechanical and electronic components. Geometry, materials, and manufacturing process are as important as function for these prototypes.

For the mechanical system, a physical prototype was hand built to match the CAD model. Once the 1st mechanical prototype was built and verified, the design was not changed. The 1st prototype ended up being the final design that went into production. Thus, the solid modeling on Inventor fulfilled both the functional and product prototype needs.

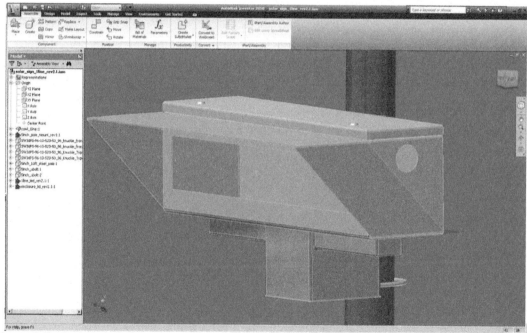

Figure 4: Early Autodesk Inventor solid model of the sign housing

Electronic PC boards were made, replacing the breadboard wiring with etched circuits. Changes could be made to these boards in hours or minutes by cutting leads on the boards and soldering in wires. However, it took days to get new boards made. As is usually the case, software changes at this stage took minutes.

The proof-of-product prototypes were installed at test customer locations to verify how they worked in real operating conditions. At this stage, much was learned about weather proofing, glare off the display and understandability of the annunciation system.

During the last design spiral, proof-of-process prototypes were used to verify the function, geometry and the manufacturing process. As noted earlier, the first physical prototype carried through to this spiral with no changes needed. Since the first prototype used the same manufacturing methods as the final product this spiral was not needed for the mechanical system. However, the next generation of the sign will use more tooling-intensive manufacturing methods such as castings and extrusions, which will result in the need for this loop in the future. As in the earlier loop, changes to the PC boards took days and changes to code minutes.

Using the spiral development process and timing the development so the three types of systems interacted at coincident times enabled Syncromatics to identify risks – potential problems - early on and make decisions that resolved them before too much commitment was made.

Syncromatics has designed the system so that any changes to the system can be made through software changes and not to the hardware and electronics, thus keeping changes quick and inexpensive.

Conclusions

Using three spirals to manage the development of the Solar Powered Shelter Signs helped Syncromatics:
- Iterate so each system could be revisited during each cycle.
- Reassess the requirements during each cycle
- Make good use of prototypes and simulations
- Enable "good enough for the moment" implementations
- Drive the level of effort by risk considerations
- Give them clear decision points in each cycle
- Each cycle provided objectives, constraints, alternatives, risks, review, and commitment to proceed.

Acknowledgments

Stephen Salazar of Syncromatics Corp., Los Angles California assisted in the development of this case study.

Autodesk® Inventor® sponsored the development of this case study.

Reinventing the See-Saw at BigToys
A Case Study for The Mechanical Design Process

Introduction

Until late in the 20th century a see-saw or teeter-totter was found in nearly every public park and school playground in America. The concept of a see-saw probably goes back before recorded history – to the first time some cave-kid put a stiff branch across a log and had his friend sit on the other end. But, with the concern for safety and with increased litigation, see-saws had all but disappeared by the beginning of the 21st century. Responding to a play void, BigToys Inc., a manufacturer of playground equipment reinvented the see-saw in 2005 as a product they call The Rock'n Cross™[i].

Figure 1: Big Toys Rock 'n Cross

Where the see-saw evolved over time, The Rock'n Cross was purposely designed in tight cooperation with customers, right down to the product's name. How BigToys went about this classic-toy redesign will be explored in light of Quality Function Deployment (QFD) and other customer-centric best practices. Thus, this case study focuses on the design of a 21st century see-saw.

The Problem: Design a new see-saw that can meet 21st Century safety concerns.
The Method: The Rock'n Cross was designed in close cooperation with children, those concerned with their safety, and other customers. The conceptual design process

loosely followed the steps found in Quality Function Deployment (QFD) and the product was developed on Autodesk® Inventor®.

Advantages/disadvantages: Focusing on customers and interacting with them takes time but results in a better product. The Rock 'n Cross has been well received by Big Toy's customers.

Background

The see-saw has been around a long time. There are Greek vase paintings of children playing on seesaws and the toy surely dates back long before then. One early depiction of it in art is in Jean-Honore Fragonard's 1755 painting, "The See-Saw" (Fig. 2). In this romantic image, the cooperative play

Figure 2: The See Saw, 1755

feature that has made the toy a favorite throughout history is clear with three people actively enjoying the experience. This "cooperative play" feature is also an important feature of the Rock 'n Cross as can be seen with the many children playing on it in Fig. 1.

Modern playgrounds began in the US around the turn of the twentieth century. These playgrounds were conceived as a respite for urban children and a place for exercise. The 1960s saw another boom in playground development, spurred by John F. Kennedy's Council on Youth Fitness. Again, the dominant purpose was exercise. These playgrounds were mostly the same: paved ground, chain-link fences, and steel swings, slides, see-saws and merry-go-rounds.

In spite of the long history and cooperative play advantages, the see-saw began to disappear from public playgrounds and schools in the United States in the 1990s. This was due to safety hazards, or at least perceived safety hazards. Historically, see-saw related injuries are caused by:

- Falling from the raised side or while standing on top of it,
- Smashing into the ground by the person on the high side when the lower person jumps off,
- Pinching injuries from fingers in the pivot mechanism,
- Hitting injuries with a person standing on the ground being struck by the see-saw either with people on it or with it pushed in motion while empty.

While the injury potential may seem great, some statistics can put them into perspective. Each year more than 215,000 American children go to emergency rooms due to playground injuries. But of these, only about 3%[ii] of the injuries are attributable to see-saws, with a majority of these due to falls on hard surfaces. This 3% compares to monkey bars (29%), slides (27%), swings (25%), and play structures (16%). Across all these injuries, falls account for 67% of them while being hit by, or hitting against equipment, is another 12%. Of the see-saw injuries, only 6.5% resulted in hospitalization compared to 14% for monkey bars and 7.5% for swings. So, even though see-saws are perceived as dangerous, relative to other playground hazards they are not a major cause of injury. With proper surfaces under them, see-saws are not very dangerous with the perception greater than the reality.

With this background, this case study focuses on how BigToys used their customers to help them redefine the see-saw. It primarily focuses on participatory design with children, even down to a contest for children to name the product.

Hearing the voice of the Customer

An important first step in any design is to learn what the customer wants, not just what engineers think they want. An oft used method to accomplish this is to use the Quality Function Deployment (QFD) method. BigToys did not formally use QFD, however they executed many of the steps in the process and so the QFD serves as a good framework for describing Big Toy's early design work on the Rock n' Cross, the see-saw for the 21st century.

The QFD method was developed in Japan in the mid-1970s and introduced in the United States in the late 1980s. Using this method, Toyota was able to reduce the costs of bringing a new car model to market by over 60 percent and to decrease the time required for its development by one-third. It achieved these results while improving the quality of the product. A recent survey of 150 U.S. companies shows that 69 percent use the QFD method and that 71 percent of these have begun using the method since 1990. A majority of companies use the method with cross-functional teams of ten or fewer members. Of the companies surveyed, 83 percent felt that the method had increased customer satisfaction and 76 percent indicated that it facilitated rational decisions.

Applying the QFD steps builds the *house of quality* shown in Fig. 3. This house-shaped diagram is built of many rooms as shown in the left half of the diagram. On the right is a house that was completed for the design of "aisle chair", a wheel chair that fits in airliner aisles. This study was completed for Boeing as part of their 787 Dreamliner development and is described in more detail in *The Mechanical Design Process*. As can be appreciated, the QFD can get complex, but here the rooms in the house are only used to describe Big Toy's effort.

Developing information in the house of quality begins with identifying *who* (Room 1) the customers are and *what* (Room 2) it is they want the product to do. In developing this information, we also determine to whom the "what" is important—*who versus what* (Room 3). Then it is important to identify how the problem is solved *now* (Room 4), in other words, what the competition is for the product being designed. This information is compared to what the customers desire—*now versus what* (Room 4 continued)—to find out where there are opportunities for an improved product. Next comes one of the more difficult steps in developing the house, determining *how* (Room 5) you are going to measure the product's ability to satisfy the customers' requirements. The *hows* consist of the engineering specifications, and their correlation to the customers' requirements is given by *whats versus hows* (Room 6). Target information—*how much* (Room 7)—is developed in the basement of the house. Finally, the interrelationship between the engineering specifications are noted in the attic of the house—*how versus how* (Room 8).

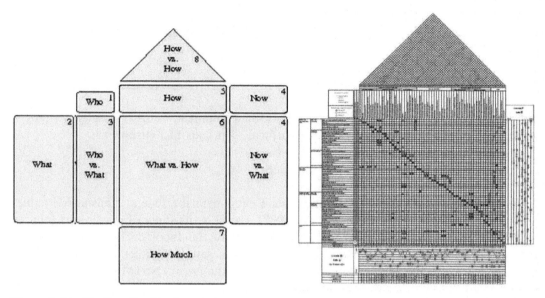

Figure 3: Quality Function Deployment

ROOM 1: IDENTIFY THE CUSTOMERS: WHO ARE THEY?

For most design situations there is more than one customer. In *The Mechanical Design Process*, a customer is defined as any person who comes in contact with the product, its manufacture, sales, and use. The customers are the "who" in the QFD diagram (Room 1). The most obvious customers for see-saws are children. But not all children play the same. The Consumer Product Safety Commission (CPSC) categorizes play ground equipment by age groups: toddlers, <2 years old; Preschool-age, 2-5 years old; School-age, 5 -12 years old; and > 12 years old. The Rock n' Cross was designed for school-aged children, but 5 year old children play differently than 12 year olds do, a difference described in the section below on Prototype Testing. Thus, the first two customers are:

- 5 year old children
- 12 year old children

The second most important customer is the playground supervisor. These are the adults who are responsible for safety and for equipment logistics (e.g. what to do if there is more demand than there is equipment). It is estimated that more than 40 percent of playground injuries are related to inadequate supervision. Since many playgrounds in the US are at schools and the supervision at schools is provided by teachers, this estimate highlights another influence on the demise of the see-saw. Namely, those responsible for playground supervision want equipment that is easy to supervise and safe to use. It is clear that the demise of see-saws is as much about supervision as it is about safety[iii]. Thus, the third customer is:

- Playground supervisors

But, they are not the ones who actually specify the equipment. This is usually done by a landscape architect, school principal or park official. Thus, another customer is:

- Playground equipment specifiers

To support the development of safe playground equipment the federal Consumer Product Safety Commission (CPSC) was created in 1972. Based on its study of playground injuries it produced Handbook 325[iv], "Public Playground Safety Handbook". This handbook, last updated in 2015, details the safety requirements for seesaws and other equipment. Also, there are at least 5 ASTM standards for playground equipment (ASTM STD F1487[v]) and surfaces, leading to another see-saw customer:

- Safety Standards and their organizations

Finally the people who install and maintain the equipment are also customers, adding one more customer to the list:

- Installation/Maintenance personnel

In developing the Rock n' Cross, BigToys kept all of these customer groups in focus through working to understand what each of them wanted in the product.

ROOM 2: DETERMINE THE CUSTOMERS' REQUIREMENTS: WHAT DO THE CUSTOMERS WANT?

The second step in QFD (Room 2) is to determine what each of the customer groups wants in a see-saw. As described in *The Mechanical Design Process* this needs to be done directly through interview, questionnaires or observations. It is too easy for engineers to think that they know what the customer wants when, indeed, they do not.

A partial list of requirements is given in Table 1. The first three customer requirements are in the voice of the children. But, supervisors also voice the need for fun and cooperative play as equipment that provides these capabilities are easier to manage. The last customer requirement is needed as some playground equipment is installed where holes can't be bored into the ground such as indoors.

Requirement	Customer(s)
Fun to Play on	Children and supervisors
Can play with my friends	Children and supervisors
Can play alone if I am the only one there	Children
Needs minimal supervision	Supervisors
Safe to use	Supervisors, Specifiers and Safety Standards
Reliable	Supervisors and maintenance
Meet ASTM Std F1487, CPSC 325	Specifiers
Inexpensive to purchase, install and maintain	Specifiers, installation and maintenance
Can be surface mounted	Installers

Table 1: Room 2 of the QFD, the Customers' Requirements

ROOM 3: DETERMINE RELATIVE IMPORTANCE OF THE REQUIREMENTS: WHO VERSUS WHAT

The next step in the QFD technique is to evaluate the importance of each of the customers' requirements (Room #3). BigToys did not do this in any formal way, but treated all the requirements as important.

STEP 4: IDENTIFY AND EVALUATE THE COMPETITION: HOW SATISFIED ARE THE CUSTOMERS NOW?

The goal in Room #4 of the QFD is to determine how the customer perceives the competition's ability to meet each of the requirements. The purpose for studying existing products is twofold: first, it creates an awareness of what already exists (the "now"), and second, it reveals opportunities to improve on what already exists. In some companies this process is called *competition benchmarking* and is a major aspect of understanding a design problem. In benchmarking, each competing product must be compared with customers' requirements (now versus what). Here we are only concerned with a subjective comparison that is based on customer opinion.

BigToys primary competition was the traditional see-saw. Although BigToys did not complete a comparison as formal as this, their opportunities become obvious using the QFD format. In Table 2, the scoring is on a scale of 1 to 5:

1 = The product does not meet the requirement at all.
2 = The product meets the requirement slightly.
3 = The product meets the requirement somewhat.
4 = The product meets the requirement mostly.
5 = The product fulfills the requirement completely.

Requirement	Traditional see-saw				
	1	2	3	4	5
Fun to Play on					●
Can play with my friends			●		
Can play alone if I am the only one there			●		
Needs minimal supervision		●			
Safe to use		●			
Reliable					●
Meet ASTM Std F1487, CPSC 325			●		
Inexpensive to purchase, install and maintain					●
Can be surface mounted					●

Table 2: Room 4 of the QFD, The competition

As can be seen, the new product should be one that does not hurt the play quality or reliability for the traditional see-saws (score = 5). The market opportunity (scores of 3 or less) is for a product that can entertain more children in a safe manner; can even be used alone, if desired; and is safe to use.

The goal of the next four steps in the QFD is to determine how to measure the customers' requirements. These four will be discussed together.

STEP 5: GENERATE ENGINEERING SPECIFICATIONS: HOW WILL THE CUSTOMERS' REQUIREMENTS BE MET?

STEP 6: RELATE CUSTOMERS' REQUIREMENTS TO ENGINEERING SPECIFICATIONS: HOW TO MEASURE WHAT?

Step 7: SET ENGINEERING SPECIFICATION TARGETS AND IMPORTANCE: HOW MUCH IS GOOD ENOUGH?

STEP 8: IDENTIFY RELATIONSHIPS BETWEEN ENGINEERING SPECIFICATIONS: HOW ARE THE HOWS DEPENDENT ON EACH OTHER?

Many small companies who produce low technology products do not feel that they have the resources to follow these QFD steps. Also, companies that design products with high human interface often feel that requirements like "fun to play on" can not be reduced to measurable engineering specifications. However, projects like the aisle chair shown in Fig. 3 give strong indication that working through these steps is worthwhile. None-the-less, BigToys did not do these QFD steps. Instead they developed concepts and built prototypes of their favorite design using these prototypes to determine, in an informal way, how well the customers' requirements were met.

Concept Development

BigToys did not start out to design a new see-saw. They contracted with an industrial designer who specialized in playground equipment and had worked closely with BigToys for many years to develop ideas for cooperative play products. He saw the demise of the traditional see-saw as an opportunity and included sketches of what was to become the Rock n' Cross in a group of ideas presented to BigToys'. Two of his sketches are shown in Fig. 4.

35

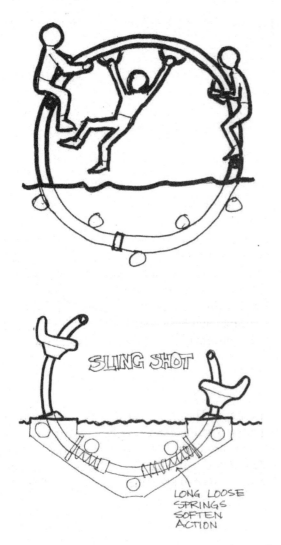

Figure 4: Sketches for what was to become the Rock n' Cross

The sketches show the original germ of the concept. In each, the tilting of the traditional see-saw has been replaced by a mechanism that moves the riders in the same circular path as is traditional, but using a rotating arc rather than a pivot. Where see-saws balance due to weight distribution, these concepts are centered using springs as shown in the right-hand, "Sling Shot" diagram. In the diagram on the left, the concept that was to become the Rock n' Cross, the springs are not shown, but are implied. Also, shown on the left-hand diagram is the cooperative play that was to become an important feature of the Rock n' Cross.

The idea of using springs to center see-saws is not new as there are units on the market such as the example in Fig. 5. Here the springs not only center the riders rotationally, they support them and provide lateral stiffness. This style of see-saw is unconstrained in its motion about all axes and in all directions. Its performance is very dependent of the weight of the riders. The Rock n' Cross on the other hand is constrained to one degree of freedom, like a traditional see-saw.

Figure 5: Spring-centered see-saw

In Table 3, the Rock n' Cross concept shown in the left-hand diagram in Fig. 4 is compared to the traditional see-saw using the customers' requirements by updating Table 2. The concept fulfils most of the requirements, at least on paper. It is only given a score of 3 on the first two requirements because these are hard to evaluate on paper. They could have been better accessed without building prototypes if the final steps of the QFD had been completed. Also, the reliability is unknown from the sketches. Finally, the concept cannot be surface mounted and it is given a low score for this measure.

Requirement	● Traditional see-saw ■ Rock n' Cross concept ▲ Final Rock n' Cross				
	1	2	3	4	5
Fun to Play on			■?		▲●
Can play with my friends			● ■?		▲
Can play alone if I am the only one there			●		■▲
Needs minimal supervision		●			■▲
Safe to use		●			■▲
Reliable			■?		▲●
Meet ASTM Std F1487, CPSC 325			●		■▲
Inexpensive to purchase, install and maintain		▲■			●
Can be surface mounted	■				▲●

Table 3. "Now" room of QFD updated with concept and final Rock n' Cross

Based on the market potential and increased safety over the competition and the other features made evident in Table 3, the Boy Toys internal steering committee selected the

Rock n' Cross concept as one idea to develop to the prototype stage. Focus during detail design was to make all the requirements score 5, with the exception of cost. Where traditional see-saws are very inexpensive, for the Rock n' Cross, cost was clearly going to be higher due to relative complexity. Purchase cost was not seen as critical and subsequent sales have proved this assumption correct.

Prototype Testing

Based on the concepts, detailed design was begun and a prototype was built out of plastic materials. During the detailed design, sufficient new functionality was developed that a patent was applied for and granted. The patent featured nine claims. Paraphrasing the first claim:

> There is an arc shaped playground device that has two parts. The first part is a fixed rail and the second a moving member with seats on the ends. The moving portion is spring loaded to a predetermined position allowing it to move to-and-fro on the rail.

This claim can be seen in Fig. 6, an assembly drawing showing (on the top) the "fixed rail", a pipe, with the springs added in the middle drawing and then finally with the "moving member" shown on the bottom.

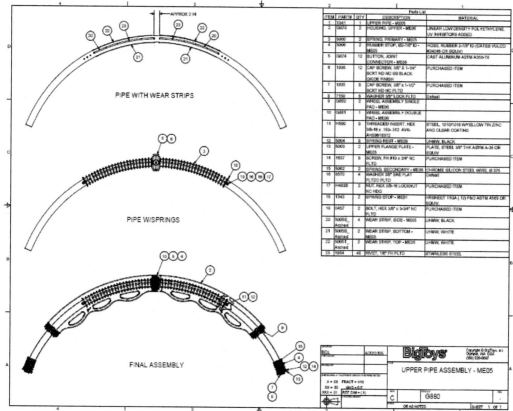

Parts List				
ITEM	PART#	QTY	DESCRIPTION	MATERIAL
1	1341	1	UPPER PIPE - ME05	
2	G870	2	HOUSING, UPPER - ME05	LINEAR LOW DENSITY POLYETHYLENE, UV INHIBITORS ADDED
3	5000	2	SPRING, PRIMARY - ME05	
4	5006	2	RUBBER STOP, O2-7/8" ID - ME05	HOSE, RUBBER 2-7/8" ID (GATES VULCO #24048 OR EQUIV)
5	G874	12	BUTTON, JOINT CONNECTOR - ME05	CAST ALUMINUM ASTM A356-T6
6	1936	12	CAP SCREW, 3/8" X 1-1/4" BCKT HD NC 88 BLACK OXIDE FINISH	PURCHASED ITEM
7	1935	8	CAP SCREW, 3/8" x 1-1/2" BCKT HD NC PLTD	PURCHASED ITEM
8	7150	8	WASHER 3/8" LOCK PLTD	Default
9	G852	2	WHEEL ASSEMBLY SINGLE PAD - ME05	
10	G851	1	WHEEL ASSEMBLY DOUBLE PAD - ME05	
11	H500	8	THREADED INSERT, HEX 3/8-16 x .180-.312 AVAIL AHS9616312	STEEL, 1010/1016 W/YELLOW TIN ZINC AND CLEAR COATING
12	5004	2	SPRING REST - ME05	UHMW, BLACK
13	5003	2	UPPER FLANGE PLATE - ME05	PLATE, STEEL 3/8" THK ASTM A-36 OR EQUIV
14	1537	8	SCREW, FH #10 x 3/4" NC PLTD	
15	5002	2	SPRING, SECONDARY - ME05	CHROME SILICON STEEL WIRE, Ø.375
16	5570	4	WASHER 3/8" SAE FLAT PLTDC PLTD	
17	H4338	2	NUT, HEX 3/8-16 LOCKNUT NC HDG	PURCHASED ITEM
18	1343	2	SPRING STOP - ME01	HRSHEET 11GA (.12) P&O ASTM A569 OR EQUIV
19	0467	2	BOLT, HEX 3/8" x 3-3/4" NC PLTD	PURCHASED ITEM
20	5005B Arched	4	WEAR STRIP, SIDE - ME05	UHMW, BLACK
21	5005B Arched	2	WEAR STRIP, BOTTOM - ME05	UHMW, WHITE
22	5005T Arched	2	WEAR STRIP, TOP - ME05	UHMW, WHITE
23	1934	48	RIVET, 1/8" FH PLTD	STAINLESS STEEL

BCL 4/20/2005 BigToys Copyright © BigToys, Inc. Olympia, WA 98501 (360) 528-9697

UPPER PIPE ASSEMBLY - ME05

C G880

Figure 6: Pipe Structure of the Rock n' Cross

Moving all the springs and guides internal to the part above the ground (as shown in Fig. 6), resolved the poor "surface mounted" score in Table 3.

The prototype was designed to test the function of the product. As pointed out in *The Mechanical Design Process*[vi] and in the case study *"Spiral Product Development at Syncromatics"* [vii] there are three reasons to develop prototypes, to test the function of the product, to test the design of the product itself (i.e. fit assemblability, etc), or to test the processes for manufacturing the product. This first prototype was aimed at testing the function with children in a controlled environment.

Figure 7: Rock n' Cross truck

One detail developed and tested during this period was the truck, Fig. 7. The truck serves two purposes. It locks together sections of the moving tubing and supports the wheels that run on the fixed pipe. In the figure, the red wheels roll on the inner steel pipe and the green, round fasteners attach a section of outer pipe to the truck. An adjacent section of pipe is attached to the truck in a similar manner, but is not shown here.

After design refinements were made based on tests on the first prototype, a refined product prototype was developed and installed on a playground at a local school. At first the children played on it, but after 1 month no one was using it. BigToys interviewed and observed the children to learn why the loss of enthusiasm for the product. They learned that:

- Five year olds play differently than twelve year olds. The younger children are more cooperative. As shown in Fig. 1, some children act as helpers to the children who are sitting on the Rock n' Cross. The idea of a helper may have been hard to see during the QFD development, but on-the-other-hand, the painting in Fig. 2 clearly shows a helper, another "customer" for this toy, so maybe the concept of cooperative play could have been developed earlier. Twelve year olds are more competitive and like the challenge and risk of hanging on the middle while other try to shake them off. In fact this has become such a feature of the Rock n' Cross that the logo for the toy has a stick man hanging on and being shaken. Compare this logo to the child hanging on the middle of Fig. 1. This change in play challenge has been a unique factor for the Rock n' Cross.

- Playground supervisors thought that supervision was much lower than that needed for the traditional see-saw.

Subsequent to these tests, some changes were made and the product finalized. During this period a naming contest was held as part of a PR campaign and a 1st grader from Maine submitted the name Rock n' Cross. She won a Rock n' Cross for her school.

Conclusions

BigToys has developed a 21st century see-saw that meets all the customer's requirements with the exception that it is more expensive than a traditional see-saw. However, with the increase in play value, safety and ease of supervision, this has had little effect on sales. The toy is a favorite of children and play ground supervisors. It has proven to foster cooperative play and safe to play on.

While BigToys did not use the QFD process, they achieved many of the QFD results. Using the QFD may have saved them some development effort.

Acknowledgements

This case study was developed with the aid of Brian Lovgren, Matt Haugh, and Tim Madeley of BigToys Inc., Olympia Washington.

Autodesk® Inventor® sponsored the development of this case study.

[i] Rock n' Cross details at the Big Toys web site

[ii] Based on 1996 Canadian statistics from the CHIRPP News, Canadian Hospitals Injury Reporting and Prevention Program, Issue 12, Nov 1997.

[iii] Communication with Donna Thompson of the National Public Playground Safety Institute

[iv] Public Playground Safety Handbook, Publication #325, U.S. Consumer Product Safety Commission, Dec 2015

[v] F1487 Standard Consumer Safety Performance Specification for Playground Equipment for Public Use

[vi] See page 118 in *The Mechanical Design Process*

[vii] The "Spiral Product Development at Syncromatics" case study

Achieving A Single Truth At Eclipse
A Case Study for *The Mechanical Design Process*

Introduction

Eclipse, Inc., one of the world's leading manufacturers of industrial burners used for heat treating, drying, curing, and industrial process heating had a problem. While they had plants on three continents and while their business was strong, they had at least three versions of the truth. If a part that could be manufactured in the U.S., Asia or Europe was needed, there were multiple CAD models for it, sometimes more than one on each continent. And, not every model in each of the areas accurately reflected the "official" drawing of the part in the system. Even worse, if a change was made to fix a problem or meet a customer's need, it was unlikely that the change would make it to all the part representations.

In 2006, Eclipse launched a project to get this problem under control using a data management system. This process is still a work in progress. This case study explores their journey, the benefits achieved, and evolving world of product data management. By studying Eclipse, it becomes clear why companies that make products sharing common parts, or companies that have multiple manufacturing facilities, need a Product Data Management (PDM) or a Product Lifecycle Management (PLM) system. Further, the case study will show how Bills of Materials (BOMs) and Engineering Change Orders (ECOs) are an integral part of these systems and how they are managed through them.

Figure 1: Eclipse burner

The Problem: Eclipse had multiple, disconnected locations creating 3-D models of the same parts. These different models did not necessarily exactly reflect the actual part.

The Method: They instituted a commercial Product Data Management System mixed with some home-grown systems to resolve the problem.

Advantages: A clear advantage of a data management system is one drawing for each part and the ability to keep it that way. Further, there is the ability to tie engineering part documentation into BOM, ECO, and business systems. This all leads to better product quality, decreased order-fulfillment time, and decreased waste. The only downside is the need for corporate commitment and time to implement the system.

Eclipse: The Problem and the Commitment to Resolve It

Eclipse, Inc. is a worldwide manufacturer of products and systems for industrial heating and drying applications. They produce a wide variety of gas and oil burners, heat exchangers, complete combustion systems, and accessories for combustion systems. Eclipse has five engineering locations and four manufacturing sites with many products common among these sites. The problem was that, by the late 1990's, parts for many of these machines had been redrawn at different plants and with different CAD systems leading to "three variations of the truth" (a quote from Eclipse product engineer Scott Stroup).

Eclipse has nearly 50 different families of products. It is primarily a "configuration to order" company with 60% of its products customized from standard parts to meet the needs of the customer. The other 40% of their business is "engineered systems" with standard parts re-engineered to meet the customers' needs. There are over 40,000 parts defined with many of the products using common parts. Using common parts is an efficient way of getting the most usefulness out of each of them and keeping inventory to a minimum. But, it also means that a single part may have to mate and work with a variety of other parts. Thus, any change in one part may not only affect one product, but many products. By structuring their products in this way, Eclipse can more easily customize products to meet the needs of the customer. This is sometimes called "variant design."

Using variants is not unique to Eclipse. For example, when ordering a new computer from a company such as Dell, you can specify one of three graphics cards, two different battery configurations, three communication options and two levels of memory. Any combination of these is a variant that is specifically tuned to your needs. Also, Volvo Trucks (and their subsidiaries, Mack Trucks in the U.S. and Renault Trucks in France) has eight wholly-owned assembly plants and nine factories owned by local interests. About 95% of the company's production capacity is located in Sweden, Belgium, Brazil and the U.S.A. Volvo Trucks estimates that of the 50,000 parts it has in its inventory, it annually supplies over 5,000 variants or different truck models specifically assembled to meet the needs of the customer.

In 2006, Eclipse decided to develop sharable documentation for key products that were sold and maintained on a global basis. The goal of this was so that a part made in any of Eclipse's plants would fit in a product made in another plant. So far key products now have a single "truth", but the process is still not complete. To understand the difficulty of such a transition, Eclipse's history needs to be explored.

As shown in the Table below, in the 1990's, Eclipse-Rockford (Eclipse's main U.S. plant) began to move from paper drawings to CAD. These were 2-D and initially on a system that no longer exists. Their facility in Holland, the facility needing the closest integration with the U.S. plant, began to convert to AutoCAD in 1996. There were over 40,000 part drawing files in Holland alone. In 2000, the U.S. office began to convert their drawings to AutoCAD, but still in 2-D. The U.S. and Dutch effort were, for the most part, independent and so a single part might have a different U.S. and European reality. In 2003, the Dutch began to use Autodesk Inventor, a solid modeling capability, and had completely converted to this system by 2005. In 2004, the U.S. began to convert to Inventor.

Year	Rockford	Europe
Pre 1995	Paper drawings of parts	
1996	SimCAD	AutoCAD (2-D, top level assemblies)
2000	Effort to get every-one on AutoCAD	
2003		Autodesk Inventor (Solid Models)
2004	Begin conversion to Inventor	
2005		Fully on Inventor
2007	75% 2D, 25% Solid models	
2009	Most standard parts and assemblies on Inventor	
2006-2009	Integrate Autodesk Vault	

These conversions further clouded the truth as:
1. Not all 2-D drawings were correct and these errors become evident when making a solid model of a part and including it in an assembly. Correcting these errors added new inconsistencies.
2. The people who modeled these parts simply needed a 3-D model to complete their assembly. They were not responsible for maintaining the official design standards. They did not need, nor did they make, a specific effort to insure that their models fully represented the piece part drawing.
3. The groups of people creating these "multiple versions of the truth" were simply creating models to fulfill their immediate need to process an order. The Product Engineering Group did not have the resources nor did they have the charter to convert all existing standard component parts into 3D models.

By 2005, things were beginning to come apart with uncontrolled part drawings. There were often three different models for the same part: a 2-D drawing in the U.S.; a solid model in the U.S. that was not validated relative to the older 2-D drawing; and a solid model in Europe. It was not that Eclipse is sloppy company. On the contrary, Eclipse is a company that has strict processes and standards. It is just that this is a hard problem to resolve and Eclipse was operating separate, stand-alone systems.

Eclipse had another common problem. In many CAD systems, the drawings are not directly tied to the Bill of Materials (BOM). Thus, when a new order was initiated, sometimes people would miss listing all the parts needed to make the product on the BOM. This error sometimes was not caught until assembly - an expensive error.

These problems led Eclipse to invest in a PDM system. Product Data Management is helping them to not only get the part and assembly drawings under control, but to streamline their product variants and their BOM system. With an integrated system of a solid models tied to a parts library, as the product is designed, the BOM can be automatically generated.

The Data Management Process

Companies that grow multi-nationally, that grow by acquisition and mergers, or that make "variant products", all face the same problems as Eclipse. And, like Eclipse, all have turned to systems that integrate the control of data files, detailed information about parts, and business systems.

This emphasis managing product information goes by many names. The two most common are PDM and PLM. Product Data Management (PDM) evolved in the 1980's to emphasize the controlling and sharing of the product data. Some began to use the term Product Life-Cycle Management (PLM) in about 2001 as a blanket term for computer systems that support the definition or authoring of product information from cradle to grave. Regardless of the name, the goal is to manage product information in forms and languages understandable by each constituency in the product life cycle—namely, the words and representations that the engineers understand are not the same as what manufacturing or service people understand. The only way to get the inherent problems under control is to move to a PDM system which is not a trivial undertaking.

On a larger scale, in the early 1990's, The Boeing Company launched an effort similar to Eclipse's. Boeing's problem partially stemmed from the fact that a wing rib on a 737, for example, might have as many as 10 different part numbers – one for each customer (e.g. United, Lufthansa, and Southwest). This created a nightmare whenever there was a product change, and it was inefficient when ordering parts. Beginning in early 1994, Boeing initiated a process improvement activity called Define and Control Airplane Configuration/Manufacturing (known as DCAC). This project was aimed at eliminating the ills of legacy practices from the 1950's – a wing rib with ten part numbers – and led to a single source of product data for each airplane and a simplified materials management method for tracking and ordering parts, scheduling production, and managing inventory for each business stream.

This project took thousands of people over a six year period. Implementation challenges such as organizational change management, new process/system introduction, data migration/conversion, and training took commitment from the top of the organization. The first efforts began with a single parts plant in 1996 followed by additional sites with more complex parts and finally, the airplane program assembly and installation in July 2000.

The ability to manage customer-specific options and leverage reusable module configurations and part definitions, led to more streamlined processes resulting from the implementation of best practices of "business object management" as opposed to processes related to coarse-grain document management practices.

Additionally, Boeing reported a reduction in part shortages following improvements in BOM accuracy and reductions in data errors, enhanced traceability and accountability processes, and reduced cycle time and quality improvements from global sharing of product information.

To understand the path taken by Eclipse and, on a much larger scale, Boeing, it is important to understand the levels[1] of product information.

- **Level 1 Data Management: Drawing Files**
  All parts are represented in CAD files. These can be 2-D, 3-D or solid models. There is no relationship between drawings other than through assembly relationships in the CAD system. If there are multiple facilities, there might be separate drawings that are not coordinated. Eclipse was doing better than this even in 2000 as they had a home grown drawing management system, which put them at Level 2.

---

[1] Levels 2-5 mimic the functions provided by the Autodesk suite of products: Level 2 = Autodesk Vault, Level 3 = Autodesk Vault Work Group, Level 4 = Autodesk Vault Collaboration and Level 4 = Autodesk Vault Professional (nee Product Stream). Other vendor's product levels are similar.

- **Level 2 Data Management: File Control**

  File control includes managing check–in and check-out, and file history management. This is the most basic level of product data management. It ensures that data is not over-written, that only one person makes a change at a time, and that there is a history of changes to each file. This is where Eclipse was in 2005 with a system that did not integrate with the CAD systems. Engineers had to check drawings in and out of a separate system. According to Mr. Stroup of Eclipse, this was not difficult. They now use Autodesk's Vault system for this level of support.

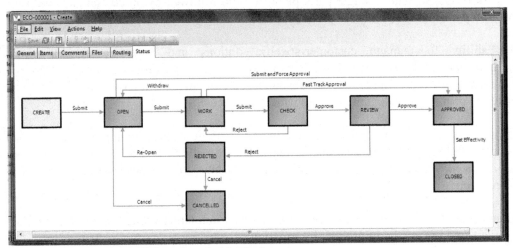

Figure 2: The process built into Autodesk's Vault system

- **Level 3 Data Management: Part Management**

  Here the focus moves from data files to the parts represented in files and how changes to parts occur. The concern here is for part revisions and releases along with life-cycle states and part properties. Fig. 2 shows the life-cycle states of a part or a change in a part as it moves through the process from its creation, through work, checking, review and finally, to approval. A Level 3 system manages this process aiding in assigning revision numbers to this release process.

  The process shown in Fig. 2 is built into Autodesk's system. This process begins to overlap processes managed by ERP systems (Enterprise Resource Planning - an integrated computer-based system used to manage financial resources, materials, and human resources). In fact, Eclipse uses their ERP system for processes such as these. As PDM and PLM mature, they are moving toward functions traditionally supplied by ERP systems and ERP is moving into traditional PDM territory.

- **Level 4 Data Management: Distributed Part Management**

  Important to Eclipse is that a data management system support the entire enterprise. This function is often called Enterprise Configuration Management, ECM, and is concerned with maintaining completeness, consistency, traceability, and accountability of the product structure across all organizations in the enterprise. This is essentially Level 3 functionality with the addition of an IT structure. For companies like Eclipse, with facilities on three continents, a Level 4 system replicates data in multiple sites to increase speed, yet the users see no difference and information consistency is maintained.

- **Level 5 Data Management: Business Object Management**

  Level 5 is a large jump to full integration of the PDM and ERP functions. Here the focus is on business objects such as BOMs and ECOs. Where Fig. 2 shows a part centric process, a business is interested in selling a product that is defined by a BOM or managing a change as defined by an ECO. To accomplish this, the part data must interface seamlessly to other business activities and systems such as Customer Relationship Management (CRM), Supply Chain Management (SCM), and Enterprise Resource Planning (ERP). This way, anyone in the organization can see the information they need in a language with which they are comfortable. This brings harmonization and synchronization to the way that companies carry out everyday business. This level is what is generally meant when the term "PLM" is used. The focus change makes Level 5 challenging as the lower levels are engineering-centric whereas, at Level 5, information is now part of many organizations and so there is no single owner.

  Eclipse has a Configurator as part of its ERP system. For a standard product (60% of sales), the sales people log the capabilities of the product and the Configurator automatically produces the BOM from the product specifications. Volvo functions in the same manner. For engineered (non-standard) products, the Eclipse engineer begins with an existing product, reengineers it in Inventor, updates the BOM to match the changes, and creates a new BOM as part of the ERP system. Eclipse has its ERP system tightly integrated with Autodesk Vault through an SQL database making this integration work for them.

The five levels are measures of data management maturity and a shift from file-centric, to part-centric, to business object-centric systems. For a company such as Eclipse to mature through these levels, there were a series of questions that needed answering:

Summary

This case study has described how Eclipse resolved the problem of "three versions of the truth" by using a Product Data Management system. This system enabled them to not only control their CAD files, but to support multiple variants, multiple languages, integrate with their ERP system, and support BOM and ECO needs.

Acknowledgements

Scott Stroup of Eclipse Inc. Rockford, Illinois; Todd Nicol, Mikel Martin and Brian Schanen of Autodesk; and Wayne Embry of Siemens PLM all helped with this case study.

Autodesk® Inventor® sponsored the development of this case study.

All Hot And Nowhere To Go At Q-Drive
A Case Study for *The Mechanical Design Process*

Introduction

John Corey had a problem. Over a period of years John had developed a thermo-acoustic cryocooler technology that enabled him to liquefy gases with simple, small-scale machines. At least they look simple on the outside and are easy to manufacture and assemble. Inside they are very sophisticated, converting electricity into a resonant acousto-thermodynamic cycle that cools to temperatures well below -200° Celsius. John built this concept into a successful company, Q-Drive, which sells cryocoolers to the military, biotech firms and others who need small, on-site sources of deep cold and liquefied gases.

John's problem was how to get rid of the heat removed. In early versions he used a convention circulating water system. But this required a water source, tubes, and often pumps, reservoir, and a separate fin-fan "radiator" greatly complicating the elegant and simple system.

This case study is about how John went from such a complex liquid coolant system to a simple heat exchanger made of eighteen interlocking extruded aluminum pieces as shown in Fig. 1. This decreased system assembly time by >90%; reduced the heat exchanger cost from $100's to $10's and weight by 50%; all with increased reliability and portability.

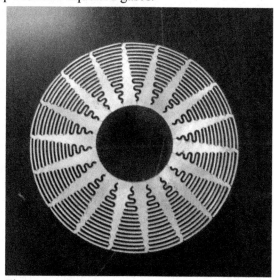

Figure 1: Heat exchanger assembly

The Problem: Develop a surface-to-air heat exchanger that doubles conventional extruded-fin performance, with no cost increase and minimal tooling expense.

The Method: John and his team used many best practices to design the new heat exchanger.

Advantages/disadvantages: The best practices allowed the designers to define the solution space, investigate existing solutions in adjacent applications and mine them for suitable approaches here, quickly extract the key features required and model the performance of the selected concept. Rapid trials of real parts enabled optimizing actual performance where obtaining similar detailed information by simulation was too slow and costly.

Background

Q-Drive coolers are useful wherever deep cold (from 50K to 200K, that is -220C to -70C) is required. They are efficient, quiet and robust. They are used for liquefying gases, for cooling instruments and superconductors, and for maintaining cold storage of medical and biological specimens. Externally, they look as shown in Fig. 2, a box with a "cold head" sticking out the right end. When mounted, this cold head is inside a vacuum-insulated flask where it cools the contents or a gas stream. The mounting flange can be seen near the base of cold head. On the end of the box at the base of the cold head is the heat exchanger that is the focus of this case study. Air is drawn into the box through the heat exchanger by a fan on the back end of the unit.

Figure 2: A Q-Drive cryocooler

The only external connection to the cooler, besides the physical connection at the flange, is an electrical cord supplying power to the unit.

Also shown in Fig. 2 are two stainless steel domes sticking out of the sides of the box. These are the heads of the Pressure-Wave Generator (PWG), part of the mechanism that converts electricity to cold. To understand what these do and why there is a need to remove heat with the heat exchanger, some basic understanding of the thermodynamic Stirling Cycle is required.

The Stirling Cycle

Thermodynamically, Q-Drive coolers work on an approximation to the Stirling cycle as shown in Fig. 3. The Stirling cycle is named for an air-cycle engine invented by Robert Stirling in 1816. The Stirling cycle has been long recognized as an efficient heat pump, but the mechanical complexity of prior embodiments has largely prevented its economical application.

At the top of the figure are idealized PV and TS diagrams. These show how heat is pumped from the cold head to the heat exchanger. In this system there is a working gas (Helium in the cryocoolers) and a regenerator. A regenerator is a porous plug with a high thermal mass that serves as a reversible heat transfer device (a stack of wire mesh in these cryocoolers).

The four labeled corners of the PV and TS diagrams define the four phases of the cycle:

- 1-2 isothermal compression: The gas is compressed at the high temperature, T_h, and heat is rejected to an external source to keep the gas temperature constant
- 2-3 constant volume heat transfer: Internal heat is transferred from the gas to the regenerator material, dropping the gas temperature to T_c.
- 3-4 isothermal expansion: The cold gas is expanded at constant temperature and heat energy is taken into the system. In the cryocooler this done at the cold head.
- 4-1 constant volume heat transfer: Internal heat is transferred from the regenerator material to the gas, raising it back to T_h.

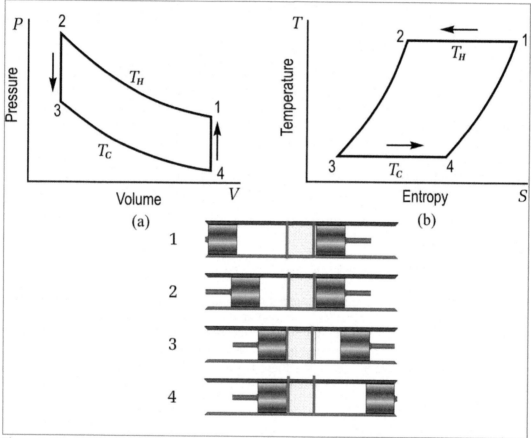

Figure 3: The Stirling Cycle

The pistons in the lower half of Fig. 3 show how this works mechanically. The pistons change the pressure of the gas and move it through the regenerator (the area between the red and blue lines in the figures). The piston on the left compresses the gas (1-2) and heat energy is removed. Then, both pistons move to the right forcing the hot gas through the regenerator (2-3) and pre-cooling it as the regenerator has a temperature gradient from hot to cold. Since the regenerator has high thermal mass it holds the heat for later in the cycle without much change in its own temperatures. The right piston then retracts, expanding the gas where heat energy is taken up (keeping the gas temperature near constant) (3-4); then both pistons move to the left, forcing the cold gas through the regenerator and re-heating it to the warm-side temperature (4-1). The two pistons' motions can be made smoothly continuous (as if crank driven) and move in a sinusoidal way, about 90 degrees out of phase with each other to approximate this ideal cycle.

Where this sequence of steps describes a refrigeration cycle (as in the cryocooler), running the cycle in the reverse order describes a Stirling engine where high-temperature heat input drives the pistons for net power output.

What has traditionally made both Stirling cryocoolers and engines difficult to design and run is the need for moving parts operating at extreme temperatures – here the cold piston on the right end of the idealized cryocooler. Q-Drive's acoustic Stirling system eliminates all moving parts, valves, and contact seals in the cold region. It has replaced these with an acoustic resonator – a tube designed with the right compliance and inertance - tuned to use the gas's own properties to function like the piston on the right in Fig. 3, 90 degrees out of phase with the left piston.   The only moving parts in a Q-Drive are the non-contact pistons of the ambient-temperature Pressure-Wave Generator, the acoustic power source for the cooling cycle.

What is in the box
The parts that make the cryocooler work are shown schematically in Fig. 4.

Pressurized helium gas is cyclically compressed and expanded relative to the mean pressure of the system, by the pistons of the Pressure-Wave Generator (PWG). These pistons are electrically driven and oscillate up and down in mechanical resonance, tuned to a specific frequency. This creates acoustic power in the form of PV work (in the wave), and pushes that work into the regenerator. Notably, there is power flow (work delivery), but the gas itself merely oscillates (no net flow) much like AC electricity delivers its power without net electron flow.

This creates a Stirling cycle by:

1) With each forward stroke of the pistons, gas moves through the hot heat exchanger (Hot HX), where heat is removed (1-2 in Fig. 3). The gas continues through the regenerator, which pre-cools it (2-3) before it reaches the cold heat exchanger (Cold HX).

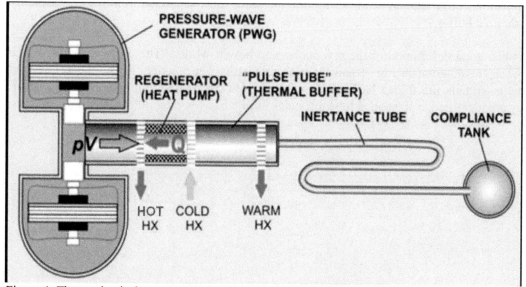

Figure 4: The mechanical components in a Schematic of a Q-Drive cooler.

2) The right end of the Q-Drive is an acoustic network tuned to a slightly different frequency than the PWG that drives it. That frequency difference means the response of flow in the network lags the driving flow from the PWG: there is a phase angle between the two local waves. The acoustic network has a thermal buffer tube, an inertance tube and a compliance tank. As gas moves toward the cold heat exchanger, gas in the acoustic network begins to move in the same direction. But, even as the driving gas stops advancing when the pistons reach their forward limits, the network's gas continues moving, driven by its own inertia in the high-speed inertance tube, and storing energy in the compliance tank (which acts like a spring). The network gas, acting like a virtual piston, draws local gas away from the cold exchanger even as the PWG's fully-forward pistons end their compressive push. This expands the gas in that area, at the cold HX. As it expands, it gathers heat from the surroundings (the area or substance to be cooled by this device).

3) The pistons soon begin withdrawing (3-4) and gas is drawn back through the regenerator. Still delayed by its inertia, the gas flow in the network slows, stops, and reverses, too, but a bit later, pushing gas back into the warm zone where the pistons have bottomed and begun their return, so the cycle - with compression at the warm zone - begins again.

The Q heat arrow in the regenerator shows how the heat flows from the cold side to the hot side (the functional purpose of the Q-Drive cooler). The source of that Q is the incoming energy at the cold heat exchanger. Together with the input pV work, the lifted heat is rejected at the hot exchanger (as in all heat pumps). A small, secondary exchanger (warm

HX) handles the frictional heating from the inertance tube, separated from the cold zone by the thermal buffer.

This is a good description of the cryocooler and how it works. The only major difference between the diagram in Fig. 4 and the real unit is that, in practice the inertance tube and compliance tank are folded back inside are built into the box that contains the cryocooler. They are identified as the coiled tube resonator in Fig. 5.

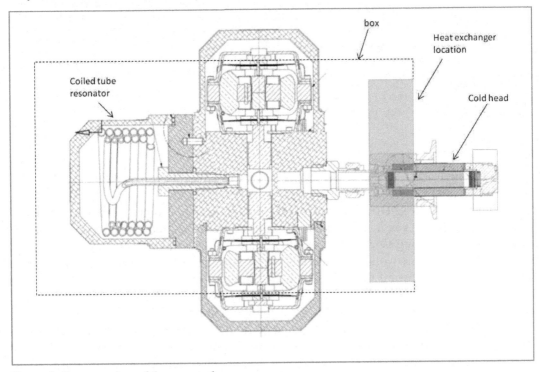

Figure 5: Cutaway view of the cryocooler

To model this device analytically requires writing equations that include mechanical, thermodynamic, acoustic, and electro-magnetic vibrations. What these equations show is that its operation and its cost are critically dependent on the performance and manufacturing of efficient heat exchangers, especially at the hot end of the regenerator (Hot HX), where the heat load is greatest.

Original Heat Exchanger

A complete cryocooler is shown in Fig. 6. The twin oscillators of the PWG are shown aligned up and down, the cold head to the right. At the base of the cold head is the original, water cooled heat exchanger. There are two visible ends of a tube there to allow circulation of cooling water. The black coil in the foreground of the figure is the electric wire supplying power to the device.

The heat exchanger shown in Fig. 6 is made from a hot-soldered attachment of cold-formed copper tubing coiled around the body of the unit, and clamped with aluminum shroud. Complete heat rejection then requires either plant-supplied water flow (non-portable installations), or a support unit with pump, reservoir, and fin-fan liquid-to-air exchanger and associated base/enclosure.

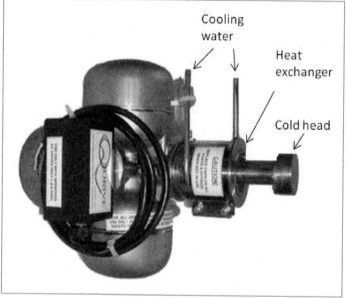

Figure 6: Cryocooler with water cooling

While the copper tube is cheap; the soldering process is difficult and costly. It must be done in a lathe to achieve even heat without excessive local temperature, and the cold head subassembly must be filled with inert gas and sealed throughout the process to prevent heat-driven oxidation of sensitive internal surfaces. Clean up of solder flux adds still more to process time and cost runs to hundreds of dollars per unit. In addition, a closed-loop liquid cooling system, purchased as a commercial component, costs over $1,000 apiece. This includes a pump, radiator, and fan along with the needed controls. Liquid-mediated heat rejection is the gold standard and has been the performance benchmark for many years as it has high heat flux capacity and very uniform temperature; but the cost, complexity, size, and reliability limits of such system limited applications where this cooler could be used.

Finding a simpler more cost effective way to remove the heat was critical to the success of the cryocooler.

The Design Process to develop a better heat exchanger

John and his team went about improving the heat exchanger in a methodical way.

Specification development

Engineers at Q-Drive first looked at the existing water-cooled system to fully define the requisite performance requirements. From this study they developed a set of measurable specifications:

Size requirements
- The outer surface of the cryocooler that must be cooled is 50 mm in diameter (base circle) and 50 mm long: a copper cylinder (the area shown in Fig. 5).
- The maximum axial thickness = 50mm.
- The maximum allowable outer diameter of the rejection components (without driving other system parts larger) is 135 mm.

Heat transfer requirements
- The heat exchanger must remove 300 Watts
- The copper surface of the hot side is 40C in 20C ambient air. Hotter rooms will drive it hotter, but these parts never get much over 60C (max operating ambient is 40C).

Power requirements
- Heat rejection should be fully portable (no secondary support services required such as a radiator or pump) – this likely means direct air cooled.
- Power requirements should be less than 10% of main cooler power.

Cost requirements
- Assembly time should be under 10 minutes and present no risk to internal components.
- Capital expense to implement (tooling, fixtures, etc.) should be under $1,000.
- Cost should be less than $100 fully assembled, in quantities of 100 assemblies per year.

They realized that the size requirements were in direct conflict with the ability to remove heat as with all heat exchangers. Also, it was evident that the cost requirement would be very challenging to achieve.

The search for concepts

They began their concept development effort by benchmarking similar devices and solutions in related applications (older style cryocoolers, computer CPU cooling, etc.). A list of alternative approaches was built (folded-foil fins, commercial heat-sink extrusion, brazed assembly of punched-plate fins, etc.). To evaluate these, they laid out the dimensional constraints (fit to base circle, allowable O.D. and length) did first-cut cost and performance estimates. It became clear that only extruded forms could hit the cost target. Extrusion tooling is inexpensive, but there is a minimum amount of aluminum extruders will run (typically 100 lbs), and sometimes a minimum number of extruded feet also. This limit encouraged the designers to look for a multi-part solution (more feet extruded per exchanger). They tried to match this with the fact that the best configurations for heat transfer are densely packed fins of brazed-plate configurations (Example in Fig. 7).

This realization led to the idea of developing a system of interlocking extruded fins in analogy to the brazed-plate configurations. This was then further refined by adding the idea of many small, identical parts that interlock to make the whole unit. If possible, a small extrusion die

Figure 7: Example of brazed-plate heat exchanger

could be used to form long lengths of material that could be cut into shorter, 50mm lengths and assembled. This thinking went from the physical constraints to developing a connected interlocking configuration of identical components.

The team began to explore different configurations that fit within the geometric constraints and model them in Excel looking at heat transfer and air flow rates. A parametric thermal performance model was built to analyze and optimize fin and segment count within the dimensional constraints. The performance challenge in designing the fin configuration lay in balancing the surface area, heat transfer coefficient (tied to spacing and flow speed) with the performance parameters of commercial fans (flow vs pressure drop). More and closer fins is more effective at any given air flow, but is worse with a real fan, if the pressure drop results in lower flow or fan stalling. Because the pressure drop here is a complex function of the main fin geometry, the minor losses of entry and exit, plus the secondary flow through the shroud box and over the main cooler body; practical analytical models were only approximate guide

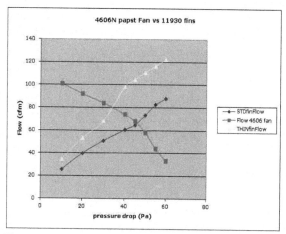

Figure 8: Flow versus pressure drop for one modeled heat exchanger

and had to be verified and refined in prototypes by testing and measuring actual performance. A typical flow versus pressure drop plot is shown in Fig. 8.

Initial proof-of-concept prototypes were made by wire EDM replication of the extrusion design. Those were tested and then modified in several iterations to reach the final parts. Only after many iterations were extrusion dies made. The goal, besides meeting the constraints was to develop parts whose connections inter-locked making assembly easy and heat transfer efficient.

The final configuration (Figs. 1 and 9) is made up of 18 identical interlocking 50mm cuts from an extrusion manufactured with a low-cost 2-inch diameter die. The fin tips interlock with notches in adjacent elements to maintain precise spacing throughout. Early prototypes did not have the tip-receiving alignment notches of the final version. These connections both help with assembly and heat transfer.

To assemble a heat exchange, Q-Drive receives the segments in a box, precut by the extruders. Silver-bearing epoxy is spread on the copper core of the cold head, while the head sits in a stand, its cold end up and with a surrounding tray at the height for supporting the segments. The segments are laid one after the other in a loose circle, interlocking, around the gluey core, and then a hose clamp is set over the whole lot and tightened to draw the segments together and onto the core. Very little finger guidance is required and the whole process takes just a few minutes (not counting glue drying time). This assembly is self aligning and epoxy guarantees good heat transfer to it,

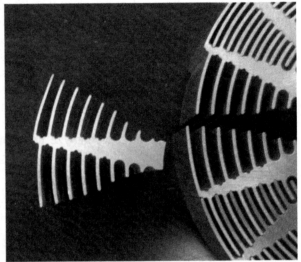

Figure 9: The final design, also see Figure 1

Conclusions

The designers carefully developed the design requirements and investigated existing solutions in adjacent applications. They used a combination of analytical models and prototypes to refine the configuration. In developing physical concepts they worked from constraints, to configuration, while driven by the connections to develop the segmented components. The system was clearly designed for easy manufacture and assembly.

Links

Q-Drive http://www.qdrive.com/UI/default.aspx

Good video showing the Q-Drive technology
http://www.youtube.com/watch?v=u8x3ihZMs7o

Author

John Corey assisted in writing this case study. John founded Q-Drive and was the chief engineer until it was acquired by Chart Industries, where John served as Vice President for Innovation.

Designing With Mushrooms At Ecovative
A Case Study for *The Mechanical Design Process*

Steelcase is the world's leader in supplying workplace products such as chairs, desks and other components for the office environment. Founded in 1912 it has a strong commitment to do the right thing for its customers, employees. This includes a concern for the environment.

Like most companies that ship products, at Steelcase packaging is critical and at the same time produces much waste. Like most others, Steelcase has cushioned their products in Styrene and Polypropylene foam packing materials for years.

©2010 Steelcase. Used with permission.

Figure 1: Ecovative's Mushroom Packaging protecting a Steelcase part

This was until they contracted with Ecovative to grow their packaging out of a fully biodegradable material made from agri-waste and mushroom roots. This Mushroom® Packaging case study is about Ecovative's entirely new process for growing products to meet Steelcase's needs. Their use of mycelium, a naturally occurring vegetative part of a fungus, as a manufacturing medium is both novel and potentially industry changing.

The Problem: Develop packaging method for Steelcase that is eco-friendly and cost competitive

The Method: Engineers at Ecovative developed a process for designing with materials that can be grown and are far more sustainable than traditional methods.

Advantages: The methods allowed the designers to consider a package design that considers the materials' lifetime beyond its primary use. This reduces environmental impact and can decrease costs, especially where legislative drivers push change. Working with new technologies requires additional time and energy, especially for early adopters.

Introduction

In 2007, Eben Bayer was an engineering major at RPI (Rensselaer Polytechnic Institute). In a senior capstone course tied to his dual degree in product design and mechanical engineering, Eben remembered his experience stocking the stove on his family's Vermont syrup farm – woodchips in the pile were clustered together, bound by white fibers. On closer inspection, these fibers inspired the introduction of industrial applications for a new biomaterial, grown from mushrooms!

Eben teamed with another RPI student, Gavin McIntyre and undertook a school project to explore the possibilities. What Eben had seen sticking the wood chips together was mycelium, the vegetative part of a fungus consisting of branching thread-like fibers. Fungal colonies of mycelium are found throughout nature, in soil and decaying plant matter, where there are high quantities of lignin and cellulose. Comparing the biological kingdom of fungi to that of plants, mycelium is similar to a mushroom's "roots," with different functions.

Eben and Gavin began to explore how to use mycelium to grow molded materials that would have commercial value. What evolved over the next several years was a process that allows them to design, manufacture, and market low embodied-energy, compostable, material that is literally grown into any custom shape and competes with plastics in performance and cost. The self-assembling bonds formed by mycelium produce this material as it grows around a substrate of regionally sourced agricultural byproducts. These bio-waste products are "glued" together by this fungal mycelium into any shape in less than a week in an indoor low-energy process, and the final material is home compostable. It is a direct, environmentally-responsible alternative to expanded polystyrene and other expanded plastics. Eben and Gavin's first efforts focused on growing insulation panels, but Mushroom® Packaging proved to be an easier product to market and sell. Mushroom Packaging represents the first time humans have capitalized on the amazing structural properties of another kingdom of biology - fungi.

Figure 2: A typical Mushroom Packaging corner block

To grow a part, in the most basic process break-down, Ecovative:

Cleans and pasteurizes the substrate, bio-waste material

Adds mycelium to it

Packs the mixture into grow-trays, simple plastic forms that are the shape of the final part

Places the grow-trays on growing racks for 3-5 days

Does nothing as the parts grow

Removes the parts and dries them, killing the mycelium and finalizing the parts integrity

A good video on this process was developed in 2011 by Energy Now, a weekly TV news magazine focused on energy issues as noted in the Links section at the end of this case study.

The key points to the manufacturing process they developed are that:

They are growing parts. This is biological additive manufacturing – a combination of bio-engineering and additive manufacturing.

They use crop wastes as a growing medium, much like the wood chips first observed by Eben. This bio-waste, like corn husks, nut shells, tree bark, or rice hulls, is available everywhere, so the process can use whatever is available in the region. A factory in Texas or China might use cotton burrs, or a factory in Virginia or Spain might use rice husks and soybean hulls. By manufacturing regionally, and using local feed stocks, the process minimizes the trucking of raw and finished materials, thus reducing the embodied energy. The goal is manufacturing within 500 miles of source of the raw materials.

Resulting products can be mulched, composted or thrown away. The material is aerobically and anaerobically compostable so it will break down quickly in a garden or home compost pile. It is 100% biodegradable and compostable, and meets ASTM standards (see Links at end of this case study).

In total, this is the ultimate sustainable material. While this process can be used to make parts for many potential products, Ecovative is currently commercially focused on replacing plastic foam made from expanded polystyrene (EPS) and polypropylene (EPP) used for packaging. According to a 2003 EPA study, expanded plastics like EPS and EPP take up about

Figure 3: Composting Mushroom Packaging

25% of landfill space by volume! This lightweight packaging is also notorious for blowing away and ending their lifecycle as litter or contributing to one of the growing ocean plastic gyres.

Ecovative has won a number of awards for the innovative technology behind Mushroom Packaging:

- 2013 Buckminster Fuller Challenge
- World Economic Forum 2011 Technology Pioneer
- Popular Science 2010 100 Best Innovations of the Year
- NREL (National Renewable Energy Lab) 2009 Industry Growth Forum Best Venture Award
- 2009 Opportunity Green Innovative Start-up Competition Winner
- 2009 New York Center for Economic Growth Rising Star Award
- Picnic Green Challenge in The Netherlands for having the greatest potential positive impact on mitigating climate change out of 200+ worldwide entries
- 2007 winner ASME iShow
- LAUNCH Innovator for 2013 Systems Challenge. LAUNCH was founded in 2010 as a strategic partnership between NASA, NIKE, the U.S. Department of State, and the U.S. Agency for International Development (USAID). Each year, LAUNCH challenges applicants to consider a different focal point for new solutions. According to LAUNCH's statement, the 2013 challenge sought "innovations that will transform the system of fabrics to one that advances equitable global economic growth, drives human prosperity and replenishes the planet's resources."

The Ecovative Design Process

As Ecovative has developed the ability to grow materials that can be molded into almost any shape, and to take advantage of this, they have also evolved a product design process. In order to keep production costs competitive, a molding system has been developed with standard design practices and rules. The steps below are how they design for replacing plastic foam packaging materials. The section after this is how this process was applied to Steelcase.

Step 1: Establish engineering requirements

Geometry needs: Size and shape of the items needing protection and the box they need to fit into. These must go along with the ability to manufacture, in Step 4.

Drop test requirements: Typical drop tests are from 30" on the corners, edges and sides of the box. The height varies with package shape, weight and item sensitivity.

Crush requirements: Packaging must redistribute crushing forces away from critical areas of the packaged item.

Vibration requirements: As goods are transported by truck or plane they undergo a spectrum of vibration frequencies. Packaging materials may need to isolate items from some frequencies.

Step 2: Benchmark the current packaging system to compare with Mushroom packaging:

Tooling cost: Packaging can be made to fit the product. Each shape made needs tooling. The tooling for Mushroom Packaging is inexpensive and can save the customer set-up costs.

Part cost: Any packaging solution may include a number of parts. The total cost of all the parts needs to be benchmarked.

Sustainability: Most packaging systems are made from EPS or EPP. The current material is a datum for comparing Mushroom Packaging (This will be done in the section of this white paper titled "Sustainability Measures").

Step 3: Design Prototype parts:

Ecovative has developed a design guide that supports engineers in determining the geometry of packaging components. It allows engineers to rapidly develop shapes that not only meet the customers' requirements, but can be easily grown. These design rules for growing packaging are similar to other Design for Manufacturability (DFM) guidelines, but are specific for grown packaging. As examples:

- No feature of a part should be less than 3/8" thick.
- All parts need a draft angle of at least 3° to easily get them out of the trays.

Besides designing the parts, the grow trays must be designed. Grow trays not only determine the shape of the part but establish the environment for the part to grow. There are also a set DFM guidelines for the grow trays. For example:

- During drying, parts contract 4-9%, thus make the grow tray 4-9% oversize on all dimensions. The percentage contraction depends on the thickness of the part and unique geometric features.

Step 4: Develop prototypes:

Since Mushroom Materials are a new and very unique packaging material, many first time customers will need to see and test prototypes in order to feel comfortable.

The prototyping process is similar to the normal production plastic grow-tray process, and it yields parts that are nearly identical to production parts.

CNC mill hard tooling from MDF (medium-density fibreboard)

Prototype grow-trays are thermoformed out of PETG on hard tooling.

Grow parts for testing. These are suitable for aesthetic approval, and testing requirements from Step1.

Step 5: Test to confirm part's function
   Conduct strength testing using drop testing, product fit testing, and other metrics as relevant
   Compare final prototype to other customer requirements (i.e.: texture, aesthetic, weight, dimensions). These become a portion of the final product's QC guidelines before product delivery.

Step 6: Build business case for customer
   Meets or exceeds engineering requirements
   Is more sustainable than the current system. See the section titled Sustainability Measures
   Is cost competitive

This entire process takes 15 -20 days. Up to 10 of these days is to grow the test parts.

Steelcase Example

Ecovative began working with Steelcase in 2009, and in June of 2010 Steelcase began using Mushroom Packaging to protect their Currency® line of ready to assemble furniture (see Figs. 1 and 4). Ecovative was able to meet and exceed Steelcase's stringent requirements for performance (ASTM and ISTA standards), environmental responsibility, and cost. When Steelcase conducted additional testing for extreme temperatures and moisture exposure, Ecovative exceeded the required standards. Based on these impressive test results, Steelcase began replacing molded and fabricated EPP parts with Mushroom Packaging for ready-to-assemble office furniture.

Steelcase had two motivations for moving to mushroom packaging. First, Steelcase is a leader in the office environments business, and is deeply focused on sustainability. They are continuously on the lookout for more environmentally responsible packaging material that would not affect damage rates. Increasing damage rates by even a fraction of a percent would un-do any good that would come from a switch to a lower embodied energy and compostable packaging material. According to the CEO of Steelcase: "Sustainability is one of today's fundamental business challenges – and our inspiration. Everyday our team works to create maximum value from our available assets and be catalysts for good."

Second, they wanted to be able to dispose the packaging responsibly, and there were no options for recycling foam in New York City, a destination for many of their products.

Throughout the development cycle, Ecovative worked closely with Steelcase's packaging engineers, sustainability experts, test labs, and marketing team to clear any doubts about bio-composite materials. Ecovative spent a year working with Steelcase's test labs, 3rd party packaging test facilities, and even brought in an expert with a PhD in cotton gin byproducts to fine tune the perfect formulation of packaging for this challenging application. Steelcase's first concern was that the packaging might soften or rot due to environmental conditions. Steelcase tested Mushroom Packaging at extreme temperatures and even simulated this high moisture condition, and it passed with no problems. Additionally, Ecovative's technology actually outperformed the EPP (expanded polypropylene) packaging that Steelcase was previously using for vibration and drop testing.

Based on these impressive test results, Steelcase began replacing molded and fabricated EPP parts with Mushroom Packaging for ready-to-assemble office furniture in June 2010. This groundbreaking solution was honored with the highest accolades in the packaging industry: The 2011 DuPont Diamond Award for Sustainability and Waste Reduction, and the Greener Package Innovator of the Year Award.

©2010 Steelcase. Used with permission.

Figure 4: Steelcase flat pack using Mushroom Packaging

Measuring Sustainability

Ecovative was obviously able to convince Steelcase that their products are more sustainable than EPP. They have been working to show this in some measurable format. The problem is that since "sustainability" is still an evolving field, and that the term "sustainable" means different things to different people, there are many "sustainability measures".

To formalize measuring sustainability, the field of Life Cycle Assessment (LCA) has begun to evolve. LCA is a technique used to assess environmental impacts associated with various stages of product life. Some LCAs are conducted for a single phase of use, while others span the product's entire lifecycle, called cradle to cradle. There are many LCA measurement tools. Ecovative works with LCA experts and software systems. Ecovative has not yet published a peer reviewed Life Cycle Assessment (LCA) for Mushroom Packaging.

In developing an LCA comparing Mushroom Packaging with Expanded Polypropylene (EPP) the engineers at Ecovative itemized the environmental impacts of each of the processes in Table 1.

Table 1: Comparison of Mushroom Packaging with Expanded Polypropylene

Expanded Polypropylene (EPP)	Mushroom Packaging
Energy Content to Produce: • The energy required to recover the oil • The energy required to process oil into PP • The energy required to make EPP (assumes in final shape)	Energy Content to Produce: • Energy gained in using bio-waste reducing needed disposal of this material • Energy required to grow mycelium (almost none) • Energy required to grow parts (almost none)
Transportation: • The energy required to transport shapes to Steelcase	Transportation: • The energy required to transport packaging to Steelcase
Disposal • Landfill	Disposal • Composted & landfill

One simple LCA tool, the Eco-indicator 99 (Eco-99), is an open tool available as a spreadsheet. The Eco-99 methodology provides a standard indicator value for a large number of frequently used materials and processes. Data have been collected in advance for most common materials and processes and a dimensionless "indicator" calculated from them in terms of points (Pt) or mili-points (mPt). The Eco-99 assessment for 1 kg of EPP is shown in Fig. 5. The resulting value, 388.5 has no meaning except in comparison with another assessment. Trying to use Eco-99 for Mushroom Packaging does not work very well, as there was no way of accounting for the use of bio-waste or for the benefits of composting. The resulting analysis yields 0.0mPt per kilogram. While this assessment clearly shows that Mushroom Packaging is far superior to EPP, it misses the key agricultural waste use and a composting disposal advantages, showing the limitations of current LCA tools.

Product or component *EPP Steelcase packaging*				Project	
Date *1/1/2014*				Author *Ullman*	
Notes and conclusions					

Production

Material or process class	Material or process detail	Units	Amount	Indicator	Result
Plastics	PP	kg	1.00	330.000	330.00
Plastics_processi	Injection moulding –	mPts		21.000	21.00
Total [mPt]					351.00

Use
Transport, energy and possible auxiliary materials

Use class	Use detail	Amount	Measure unit	Indicator	Result
Transport	Truck 16t	tkm	1	34.000	34.00
Total [mPt]					34.00

Disposal
Disposal processes for each material type

Disposal class	Disposal detail	Amount	Measure unit	Indicator	Result
Landfill	Landfill PP	kg	1	3.500	3.50
Total [mPt]					3.50
Total [mPt] (all					388.50

Figure 5: Eco-99 assessment of EPP

Conclusions

Are mushrooms the new plastics? Ecovative has begun what might be a materials revolution giving designers another material to work with in developing sustainable products. Steelcase is one of Ecovative's first customers and has continued its use of Mushroom Packaging.

Links

Good video about Ecovative Design from Forbes is at:
https://www.youtube.com/watch?v=snIvRhmsUZQ

TED talks by Ecovative Design founders
- Eben Bayer: Are Mushrooms the new plastic?, TED Presentation, July 2010
  http://www.ted.com/talks/eben_bayer_are_mushrooms_the_new_plastic.html

- Gavin McIntyre: Bio-Connectivity, TEDxPresidio,
  http://www.youtube.com/watch?v=zqG9ge_VxA4

- Sam Harrington: Innovate: Ultra-Rapid Renewables, TedxFGrandRapids,
  http://www.youtube.com/watch?v=GWJMG_Ddw8Q

- Eben Bayer: Drinking Trees, TEDxAlcatraz, 2010
  http://www.youtube.com/watch?v=VTsH8qgIb80

ASTM Standards
ASTM D6400, Standard Specification for Compostable Plastics,
http://www.astm.org/DATABASE.CART/HISTORICAL/D6400-04.htm

ASTM D5338, Standard Test Method for Determining Aerobic Biodegradation of Plastic Materials Under Controlled Composting, http://www.astm.org/Standards/D5338.htm

D5210, Standard Test Method for Determining the Anaerobic Biodegradation of Plastic Materials in the Presence of Municipal Sewage Sludge,
http://www.astm.org/Standards/D5210.htm

Sustainability tools
The Eco-indicator 99: A Damage Oriented Method for Life Cycle Assessment (EI99), http://www.slideshare.net/guest05414e/ei99-manual, Available as a spread sheet template that has been "Americanized" for student use at www.davidullman.com/MDP5/Templates

Others Links
FedEx: Testing Packaged Products Weighing up to 150lbs.
http://www.fedex.com/us/services/pdf/PKG_Testing_Under150Lbs.pdf

DuPont Diamond Award for Sustainability and Waste Reduction.
http://www2.dupont.com/Packaging_Resins/en_US/whats_new/23rd_packaging_awards_wi
nners.html

Acknowledgements

Stephen Nock and Eben Bayer of Ecovative Design LLC, Troy NewYork assisted in writing this case study.

Designing a Hybrid Car at BMW
A Case Study for *The Mechanical Design Process*

Introduction

In an effort to reduce CO_2 emissions and develop a distinctive electric driving experience, BMW initiated a program to develop a hybrid version of their 5-Series, high end car. They needed to develop a concept that on the one hand realized the fuel economy potential of hybrid technology, and on the other, offered typical powertrain characteristics and drivability. Further, they wanted an architecture that would allow them flexibility to evolve more sophisticated systems, scaling from a mild-hybrid to plug-in-hybrid.

This case study focuses on the development of drivetrain for the ActiveHybrid 5, Fig. 1.

BMW's entry into the luxury hybrid market was the model Active Hybrid 6 equipped with a power split drivetrain.

This case study describes one of the tools to explore the many hybrid architectures and some of the reasoning behind the choices made for the ActiveHybrid 5. It focuses on the use of Design Decision Matrices (DSMs) and similar tools to manage the functions and components that make up a hybrid system.

Figure 1: 2012 BMW ActiveHybrid 5

The Problem: Develop an architecture for the ActiveHybrid 5 line of drive trains.

The Method: Engineers at BMW use Multiple Domain Matrix (MDM), Function-Design Structure Matrices (F-DSM), Design Mapping Matrices (DMM) and Component-Design Structure Matrices (C-DSM) to explore the space of hybrid architectures.

Advantages: The methods allowed the designers to derive and evaluate architectures in order to develop a new concept that best achieved the goals for the new Active Hybrid 5. It helped them to frame a solution space; study available competitors' architectures; derive, analyze and evaluate specific solutions; and funnel the whole solution space, detailing promising concepts further.

Background

Traditional automobile architectures are changing. Automobile architecture, like building architecture, refers to basic ways that the functions are fulfilled by physical components and assemblies. Traditionally, automobiles have been built with an internal combustion engine (ICE) to provide power to a transmission that converts the power and delivers it to a differential that distributes the power to the wheels. The mix of functions (verbs) and components (nouns) in the previous sentence describes the traditional powertrain architecture which has remained virtually unchanged since early in the twentieth century.

BMW began to experiment with electric cars in the 1960s resulting in the BMW 1602 electric demonstrator built to move dignitaries at the 1972 Munich Olympic Games. They have undertaken periodic electric car experiments since that time and moved to serious hybrid car research and development in the 1990s. Their goal was to not only be responsive to customer interests, but to reduce CO_2 emissions and increase efficiency.

BMW began a fleet wide Efficient Dynamics strategy in 2007 to incorporate improvements in aerodynamics, rolling resistance, light weight construction, intelligent control systems and improved combustion engine technology. Part of this strategy was the addition of hybrid models with electric motors and batteries added to the traditional powertrain components. The hybrid functions and components opened up the possibility for many new architectural options and added significant complexity to the systems.

BMW was responding not only to customer demands, but also increased government standards both in the US and the EU. Cars are responsible for around 17% of emissions of CO_2 in the US and 12% in the EU. In 2007 the BMW CO_2 fleet emissions average was 158.7grams/km (255 g/mile) improving to 132g/km (212 g/mile) by 2012. The EU has specified that the fleet average to be achieved by all new cars is 130 grams of CO_2 per kilometer (g/km) (209 g/mile) by 2015, 95g/km (152 g/mile) by 2020. For BMW, these targets represent reductions of 18% by 2020. In the US, the EPA will require 144g/mile (89g/km) for cars and 203g/mile (126 g/km) for light trucks by model year 2025.

As part of their Efficient Dynamics strategy, and to optimize their learning how to best evolve their current fleet using electric hybrid technology, engineering design teams within the company operated semi-independently in realizing a production family of models. Each team could explore a range of architectures. This case study focuses on the effort on the BMW 5-Series hybrid team.

Automobile architectures

For automobiles there are four broad classes of architectures:
- ICE - Internal Combustions Engine, the traditional powertrain
- HEV - Hybrid Electric Vehicles, where the ICE is supplemented by an electric motor and batteries. HEVs are the focus of this case study.
- PHEV - Plug-in Hybrid Electric Vehicles, the batteries can also be charged by plugging into the grid.
- BEV - Battery Electric Vehicles, all electric with no ICE.

HEVs were first commercially available as an alternative to the conventional ICE powertrain in 1898 when Dr. Ferdinand Porsche developed the 'Lohner-Porsche' featuring electric hub motors. It was in production from 1900-1905. The different types of architectures are compared in Table 1 using a decision matrix with HEV as the datum.

In modern times, initial HEVs featured small electric systems that interact with the ICE. In these systems, battery charging only occurs through the ICE or regenerative braking (converting the momentum of the car to electricity by using the motor as a generator during braking). Primary amongst these HEVs, the Toyota Prius, introduced in Japan in 1997 and worldwide in 1999, has sold over 9 million units. Prius' were later updated to be also available as a PHEV where the batteries can be recharged from the grid. The Prius PHEV has an all-electric EPA estimated range of 13 km (11 mi). The Chevy Volt and its European cousin, the Ampera are PHEVs. As of mid 2017 they have combined global sales of over 120,000 units since their introduction in 2010.

Criteria	ICE	HEV (Reference)	PHEV	BEV
Electric Driving Range	NA	O	+	++
Total Range	+	O	O	--
Operating costs	-	O	+	++
Tank to Wheel Emissions	-	O	+	++
Tank to Wheel Efficiency	-	O	eDriving: ++ Total : +	++
Refueling Duration: Electric	NA	NA	-	--
Gasoline/Diesel	O	O	O	NA
Vehicle Weight	+	O	-	-
Manufacturing Costs	++	O	-	--
Commercial Risk (Battery-Tech. Maturity, Service Costs)	++	O	--	--
Ecological Image/ Possible Perks	--	O	+	++
Political Support	-	O	+	++

++ Very Advantageous; + Some Advantages; o Average; - Some Disadvantages; - - Many Disadvantages

Table 1: Comparison of classes of automobile architecture

The best selling high end BEV are made by Tesla. The Roadster, introduced in 2008 and discontinued in 2012, sold about 3000 units in its lifetime. Since then, with new models, Tesla has sold over 250,000 BEVs by mid 2017.

HEV Major Architectures

There are many ways to configure an HEV. The 5-Series design team first defined the different modes of operation needed for a hybrid system. In each, there is an ICE, a high voltage battery for storing propulsive energy and at least one electric motor connected through a controller to the high voltage batteries. This list will be refined later in the case study.

The HEV modes of operation defined by the team are:

ICE engine Start-Stop – The engine start-stop function is a basic function found in all hybrid vehicle concepts. As soon as the hybrid control system senses that the vehicle will come to a complete stop, for example at a traffic light, the engine will shut off to prevent idling. The ICE must be restarted quickly and smoothly by means of an electrical motor or starter-generator as soon as there is a power requirement.

Regenerative Braking – The term "regenerative braking" refers to capturing the braking energy that would normally be lost to friction and heat in conventional car brakes. Brake energy recuperation is achieved by using the electric traction motor as a generator that serves slow the vehicle and supply that energy to the batteries.

Power Boost – When the automobile requires acceleration power beyond what the ICE can deliver, the electric motor(s) provide additional torque to the wheels known as boosting. Power boosting situations also include driving on inclines or towing use cases. In this mode, the battery charge is depleted and delivered through the electric motors as an additional source power.

Load level Increase – Load level increase or generative mode allows the ICE to deliver some of its excess power to generate electricity for storage in the high voltage batteries. Also, since running at a more optimal setting, this increases the efficiency of the ICE itself.

Electric Driving – Electric driving is achieved by using electric energy stored in the high voltage battery to power electric motors connected to the wheels. During the electric driving mode, the combustion engine is decoupled from the powertrain and in most cases shut down. Optionally, the ICE can be used to generate electricity while decoupled from the wheels.

External Battery Charging – External battery charging (i.e. plugging into the wall or a charging station) differentiates plug-in hybrid (PHEV) concepts from all other hybrid vehicle (HEV) concepts.

There are many architectures that can provide these modes of operation. To explore these, the engineers first defined a set of icons representing the components as shown in Fig. 2. These will be used as the potential architectures are described.

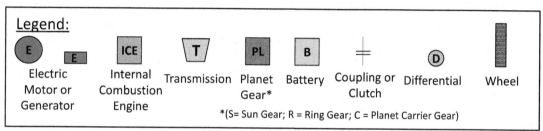

Figure 2: The components used in an HEV

These components can be used to design many thousands of different systems. Useful architectures can be organized in a tree structure (Fig. 3) with the six branches described in detail below.

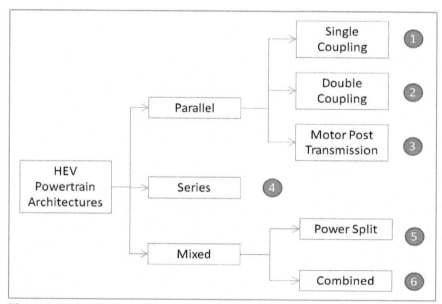

Figure 3: Major HEV architecture classification

A "parallel HEV" maintains a mechanical linkage between the internal combustion engine and the wheels, the electric motor supports the ICE (which is still the primary method of propulsion). Thus, the ICE and the electric motor run "in parallel." Parallel hybrids offer the broadest range of architectural configurations. There are over 4000 different configurations of major components in parallel HEV using the methods discussed in the next section.

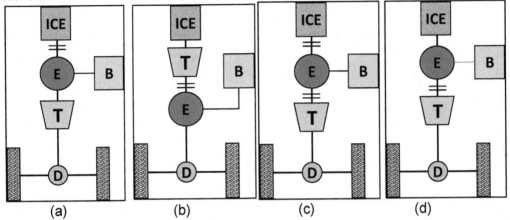

Figure 4: Parallel HEVs: (a) Single clutch - Integrated Motor Assist (b) Single clutch, (c) Double clutch, and (d) Post Transmission

The simplest HEV uses the parallel – single clutch architecture, Figs. 4(a) and 4(b). Here, if there is no power to the electric motor, the car's drive train operates in a traditional manner. Add power to the electric motor from the batteries and then both the ICE and electric motor power the wheels. Fig 4(a) is typical of mild hybrid systems such as the Honda Insight and the BMW Active Hybrid 7 that are designed to start and assist the ICE, but are not designed for electric only driving. Systems such as shown in Fig 4(b) can open the clutch to operate as an electric car, but require a separate electric motor system to start the ICE. One advantage of this configuration is the possibility to uncouple both transmission and ICE and thus reduce drag torque to maximize energy recovery during regenerative braking.

The parallel – double clutch, Fig 4(c), is similar to the prior example, except there is a second clutch between the electric motor and the transmission. By uncoupling the transmission, the ICE can drive the motor as a generator to recharge the battery. In some double clutch designs, the electric motor can also serve as a starter for the ICE. This is the system that ended up being used in the ActiveHybrid 5. More details on it will be developed in the next section.

The parallel – motor post transmission, Fig 4(d) uses the electric motor to directly power the drive train without use of the transmission. This configuration facilitates keeping existing motor-transmission conventional placement intact and adding the electric motor component outside of already established modules.

A "series HEV" completely severs the mechanical link between the ICE and the wheels. Rather, the engine serves only to produce electricity.

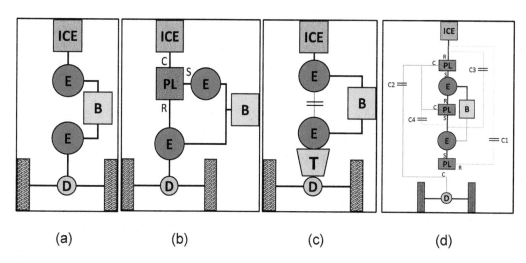

Figure 5: Series and Combined HEVs: (a) Series, (b) Power Split, and (c) Combined (d) Two-mode power split

The pure "series" configuration, Fig. 5(a) is where the ICE and an electric motor/ generator system are in series with each other. Here, all the power produced by the ICE is converted to electricity and the car is essentially an electric car. Fisker's Karma brought this architecture to mass production in 2012 and coined the term Electric Range Extended Vehicle (EREV) to emphasize it always drives the wheels electrically.

The "power split" architecture made popular by the Toyota Prius, Fig. 5(b), uses a planetary gear to channel the ICE produced energy between driving the wheels mechanically and generating electricity. In the diagram, the ICE is connected to the planetary's planet carrier, "C"; a motor/generator is connected to the sun,"S"; and the second motor/generator connected to the drive shaft to the ring gear, "R". Here a varying part of the ICE power is transferred mechanically to the wheels with the rest converted to electricity and transferred electrically. This architecture enables the car to operate the ICE at optimal efficiency. The driver may recognize driving conditions in which the ICE has a constant engine speed, but the car continues to accelerate.

Some vehicles use systems that are a mix between series and parallel architectures and are referred to as "combined". In Fig 5(c) the system can be run as a pure series or a parallel HEV by simply adding a coupling component. While there are no commercially produced combined hybrids, there are several concept cars that have combined hybrid architectures. The series mode is used for the long distance electric driving and the parallel mode is used for when the battery is depleted.

Finally, other variations of hybrids exist. Fig 5(d) shows a diagram of a two-mode power split hybrid used in the BMW Active Hybrid X6. The combination of four clutches, three planet gears and two electric motors allow for various control strategies for a variety of power delivery combinations to the drive train. This hybrid drive also allows for uncoupled electric motor use for charging the batteries, brake energy recuperation and uncoupled starting of the ICE.

BMW Series 5 Down-Select

Based on their knowledge of the Active Hybrid 7 series and other HEV systems, the engineers needed to choose an architecture and refine it. Their major considerations were:

Needed electric-only driving pattern
HEVs can be designed so that: 1) the vehicle is only assisted by the electric motor but not for pure electric driving (known as a 'mild hybrid' – such as the Honda Insight), 2) the vehicle can drive electric for short distances of a few miles for inner-city speeds and accelerations (known as a 'full hybrid' – such as the Ford Escape), or for managing typical distances such as the daily commute to work by pure electric drive (in this case a 'Plug-in hybrid' is required – such as the Chevrolet Volt). The team decided to take the middle ground and design a system that can run a few miles on pure electric and be categorized as a full hybrid, but not require external charging.

Ability to integrate in existing systems
The 6-cylinder N55 turbocharged direct injection straight-6 DOHC ICE with the 8-speed automatic transmission is the basis of the ActiveHybrid 5. The aim was to make all necessary adaptations in this well known package for integration of the electric motor. The engineers decided to stick closely to the mature and reliable transmission and fit the electrical motor in the space of the conventional torque converter within the transmission.

Easy ICE start-up capability
The design team placed extra emphasis on a smooth transition between electric driving and combined 'engine plus motor' modes of operation to ensure optimal electric vehicle performance. Excellent comfort and response characteristics during start of the combustion engine were also an area of focus. To achieve this, a belt driven starter unit connects to the

combustion engine powered separately by a 12V battery with the main function of ICE start up. This ensures that the vehicle can start from rest in pure electric driving mode, and then comfortably start-up and couple in the ICE when power demands require it.

Platform for future development
The lithium-ion battery (developed and produced in-house), the power electronics and the electric motor result in an electric powertrain system which is connected via the electronics of the combustion engine to give one harmonious unit. This can serve as a basis for future development.

The Active Hybrid 5 design team chose a parallel, double clutch system (Fig 4(c)) as the one that best met the requirements. It was shown to be the most affordable solution with the greatest customer benefits. It is comparably cost effective to produce, since it has the least modification to the existing system and has just one electric motor that could integrate within the transmission module of the vehicle, in line with the engine. Further, it is a good platform for future development with its modular architecture.

A more detailed diagram of the system is in Fig. 6. Here:
- K0 = Separating clutch between the ICE and transmission (AT) and the integrated electric motor (EM)
- IAE = Integrated start-up element
- EME = High-voltage power electronics with integrated control unit
- DC/DC = Voltage converter 317V – 12V
- SGR = 12V Starter unit in the belt drive
- EPS = Electric power-steering pump
- ELUP = Electric pump for brake servo

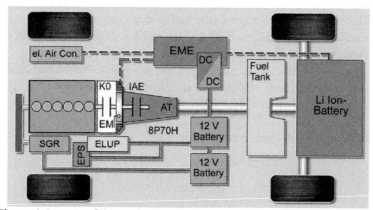

Figure 6: Layout of the Series-5 Hybrid

The design of the Active Hybrid 5 had to decouple several conventional subsystems normally tied directly to the ICE through belt systems to make them operate electrically since the combustion engine does not power the drive train during electric driving. The pump for the brake servo, the power-steering and air conditioning systems are examples of these architecture deviations from conventional ICE cars that are normally tied by belt systems that run directly from an 'always on' ICE. By creating electrically driven subcomponents, the functionality of these systems continue to remain on-line when the ICE is shut off completely but the vehicle continues to run in electric mode.

Note that the system has a high voltage and low voltage system. The high voltage batteries (an Lithium-Ion Battery) power the motor, the control unit and the air conditioning system that is needed by both the batteries and the interior cockpit. The control unit (labeled as EME in Fig. 6) also has a DC/DC converter to interface with the conventional low voltage components and electronics.

The double clutch configuration allows the car to decouple the ICE from the electric motor by leaving the K0 clutch open when driving in electric only mode or during brake energy recuperation. It also allows for the electric motor to decouple from the transmission and be used as a generator for the high voltage system (317v) using engine power when the car is stationary and the controller senses the battery is depleted.

When battery power is available, an electric start from rest is always preferred by the hybrid control unit, delivering the highest traction motor torque almost instantaneously to the wheels. This is achieved by leaving the K0 clutch open and closing the integrated starting element (IAE) clutch. In this coupling configuration the electric motor and high voltage battery powers the vehicle.

When additional power beyond that of the electric motor is required, the belt driven starter generator (SGR) starts the combustion engine (the belt is depicted in the diagram as the thick line on the left connecting the ICE and the SGR). This SGR is found standard in all 5 series and delivers the motor start-stop function, along with some minor recharging of the 12V system - like in most so called 'micro' hybrids. Keeping the same engine and SGR package from the conventional 5-Series was selected over adding a second electric motor. This decision was key for capitalizing on production economies of scale.

Fig. 7 shows the transmission with the electric motor.

Figure 7: The 8P70H transmission with the Electric Motor.

In the next section we further explore how the engineers detailed this configuration.

The Multiple Domain Matrix

The goal of this section is to demonstrate one of the tools used to explore the functional and component relationships – the architecture of the system. What is presented here is a greatly simplified version of what was considered at BMW.

During the development of the Active Hybrid 5, engineers studied existing HEV architectures. The Active Hybrid X6 was BMW's first hybrid car to go into production (2008) featuring a highly capable, yet very complex 'two-mode' power split hybrid architecture developed in cooperation with GM and Mercedes Benz, Fig. 5(d). The X6 saw small production volumes and its development formed the basis for further hybrid development.

The Active Hybrid X6 was followed by the Active Hybrid 7 series that entered production in 2009 and was fully designed at BMW. The Active Hybrid 7 series architecture moved away from the complexity of the X6, focusing primarily on the basic hybrid system functionality of a mild hybrid Fig 4(a).

1	Store Fuel
2	Store Electric Energy
3	Convert Fuel into Mechanical Energy
4	Convert Mechanical into Electrical Energy
5	Convert Electrical into Mechanical Energy
6	Deliver (Recover) torque to (from) wheels
7	Convert Moment transferred (mechanical)
8	Equate Rotation
10	Couple/Uncouple Moment
11	Release Energy as Heat to the Environment
12	Transfer Heat (to Cooling system)
13	Transfer Moment to (from) the road
14	Slow or Stop Vehicle (recovering energy)
15	Slow or Stop Vehicle (using friction)
16	Control Energy Flow
18	Consume El. Energy for Auto Accessory OPS
19	Consume Mech. Energy for Engine Accessory

Figure 8: Functions identified for components of an Integrated Motor Assist HEV.

In developing the Active Hybrid 5, developers wanted the simplicity of the Active Hybrid 7, but with the powertrain flexibility offered by the Active Hybrid X6. The development team turned to the Multiple Domain Matrix (MDM) tool to help map the fundamental relationships amongst and between hybrid functions and components. The tool helped in analysis of architectures in a standardized way, in what already was a large field of design possibilities. The MDM is built from a Functional Design Structural Matrix (F-DSM), a Design Mapping Matrix (DMM) and a Component Design Structure Matrix (C-DSM).

The first step is to fully understand the relationships among the functions using a Functional-Design Structural Matrix (F-DSM). The detailed functions for a single clutch parallel architecture hybrid are shown in Fig. 8. The architecture can be visualized in Fig. 4a. Each function was entered into the rows and columns of a Design Structure Matrix (DSM) and the inputs and output dependencies where noted with a "1" as shown in Fig. 9.

	1	2	3	4	5	6	7	8	9	10	11	12	13	14	15	16	17
1 Store Fuel	■		1														
2 Store Electric Energy		■									1				1	1	
3 Convert Fuel into Mechanical Energy			■	1							1						1
4 Convert Mechanical into Electrical Energy				■							1				1		
5 Convert Electrical into Mechanical Energy		1			■				1		1				1		
6 Deliver (Recover) torque to (from) wheels						■	1					1	1	1			
7 Convert Torque transferred (mechanical)						1	■	1	1			1	1				
8 Equate Rotation							1	■				1	1				
9 Couple/Uncouple Torque					1		1		■				1				
10 Release Energy as Heat to the Environment										■							
11 Transfer Heat (to Cooling system)										1	■						
12 Transfer Torque to (from) the road						1	1	1				■			1		
13 Slow or Stop Vehicle (recovering energy)													■		1		
14 Slow or Stop Vehicle (using friction)						1				1		1		■			
15 Control Energy Flow	1		1	1							1		1		■		1
16 Consume El. Energy for Auto Accessory OPS																■	
17 Consume Mech. Energy for Engine Accessory																	■

Figure 9: Functions identified for the Active Hybrid 7 series architecture.

A key feature of this functional DSM is the ability to see, for each function, what other functions it is dependent on (in the columns) and which functions are dependent on it (in the rows). Fig. 10 shows the basic logic scheme for the Functional DSM. The intersection of each functional element with itself is blackened on the diagonal of the matrix. Elements along the row are outputs of the function, whereas inputs are in the columns just as the arrows in Fig. 10 indicate. BMW studied many potential DSMs with Fig. 9 just a representative example.

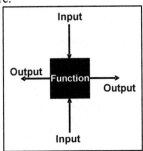

Figure 10: Functions DSM logic.

To make the best use of the functional DSM in designing architecture, it can be expanded into a DMM (Design Mapping Matrix). This allows the development of the components and their relationship with the functions. The "components" may be systems or individual parts.

The DMM maps components to functions. In some cases this may be 1-to-1 and the DMM will result in a square matrix, but this is not usually the case as some functions require many components and some components provide many functions. Fig. 11 is a DMM for the active hybrid 7 series. The functions head the columns, and the components the rows.

#	Component	1 Store Fuel	2 Store Electric Energy	3 Convert Fuel into Mechanical Energy	4 Convert Mechanical into Electrical Energy	5 Convert Electrical into Mechanical Energy	6 Deliver (Recover) torque to (from) wheels	7 Convert Torque transferred (mechanical)	8 Equate Rotation	9 Couple/Uncouple Torque	10 Release Energy as Heat to the Environment	11 Transfer Heat (to Cooling system)	12 Transfer Torque to (from) the road	13 Slow or Stop Vehicle (recovering energy)	14 Slow or Stop Vehicle (using friction)	15 Control Energy Flow	16 Consume El. Energy for Auto Accessory OPS	17 Consume Mech. Energy for Engine Accessory
18	Fuel Tank	1																
19	High Voltage Battery		1									1						
20	Internal Combustion Engine			1								1						
21	E-Motor/Generator1				1	1						1		1				
22	Transmission							1										
23	Differential Gear						1	1	1									
24	Clutch Direct Coupling1									1								
25	Cooling System										1							
26	Wheels												1	1	1			
27	Brake-system										1				1			
28	Power Electronics/Inverter											1				1		
29	Additional Electric Accessories																1	
30	Mechanical Accessories																	1

Figure 11: The initial DMM for the Active Hybrid 7 series architecture.

Note that the 'E Motor/Generator1, component # 21, serves multiple functions. It can 'convert mechanical to electrical energy' when in generative mode, 'convert electrical into mechanical energy' when working as a motor, and can be used to 'slow or stop the vehicle'.

Additionally, the motor releases heat to a cooling system. Looking at the cooling function in column 11 we see that components 19, 20, 21 and 28 all require cooling.

Finally, the engineers used a Components-Design Structure Matrix (C-DSM) to determine the final configuration of components shown in Fig. 12. The C-DSM has the components in both the rows and the columns. It is used to help understand the physical connection between components. Since there is no sense of one component preceding another, the C-DSM is symmetrical. For example, the Fuel Tank must be physically connected to the internal combustion engine and vise-versa. The location of the "1"s gives a good indication of potential component architecture.

		18	19	20	21	22	23	24	25	26	27	28	29	30
18	Fuel Tank	■		1										
19	High Voltage Battery		■						1			1		
20	Internal Combustion Engine	1		■	1				1					1
21	E-Motor/Generator1			1	■			1	1			1		
22	Transmission					■	1	1						
23	Differential Gear				1	■				1				
24	Clutch Direct Coupling1				1	1	■							
25	Cooling System		1	1	1				■					
26	Wheels						1			■	1			
27	Brake-system									1	■			
28	Power Electronics/Inverter		1		1				1			■	1	
29	Additional Electric Accessories											1	■	
30	Mechanical Accessories			1										■
		18	19	20	21	22	23	24	25	26	27	28	29	30

Figure 12: Component DSM

The F-DSM, the DMM and the C-DSM were all be put together as a platform for making architecture decisions (often called the Multiple Domain Matrix (MDM)). Fig.13 shows all the pieces together, with the DMM updated to the DSM reordering. It has been assumed here that function #9, Sense Torque Need, is part of the controller. It is easy now to manage the architecture, add additional functions or components, and group functions within components.

For example, the function "convert fuel in mechanical energy" (performed by the ICE) is depicted in the middle node as taking inputs from the function "store fuel" (function of the fuel tank) to provide an output to the function "convert mechanical energy to electrical energy" (performed by the electric motor). The usage of fuel by the ICE can only occur in one direction. Other components such as the electric motor display a bi-directional energy flow in converting from mechanical energy to electrical energy and vice versa.

The value of the MDM vehicle architecture representation between the components and functions domain lies in the systematic determination of differences and similarities of various structural configurations.

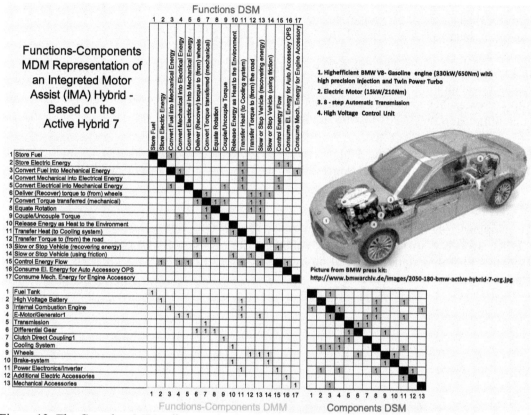

Figure 13: The Complete MDM diagram for the Active Hybrid 7 series architecture

Conclusions

There are many hybrid drive architectures that can be used in a high performance car of today. DSM and MDM diagrams enabled the BMW team to explore this solution space. The methods used and the resulting system both offer platforms for future developments and architecture evolution.

Links

C. Gorbea Diaz, "Vehicle Architecture and Lifecycle Cost Analysis in a New Age of Architectural Competition", Doktor-Ingenieurs Dissertation, University of Munich, 2011 (in English). https://mediatum.ub.tum.de/node?id=1108157

C.O. Griebel et al, The Full-Hybrid Powertrain of the new BMW ActiveHybrid 5, Aachen Colloquium Automobile and Engine Technology, 2011.
http://www.aachener-kolloquium.de/wp-content/uploads/2014/02/delayed_2011_B5.2_Griebel_BMW.pdf

C. Luttermann et.al, The Full-Hybrid Powertrain of the New BMW Active Hybrid 5, Aachen colloquium China, BMW Group, 2011
http://www.aachener-kolloquium.de/china/wp-content/uploads/sites/2/2014/04/04 Luttermann.pdf

Grewe, T., Conlon, B., and Holmes, A., "Defining the General Motors 2-Mode Hybrid Transmission," SAE Technical Paper 2007-01-0273, 2007, doi:10.4271/2007-01-0273.

Acknowledgements

Development of this case study was guided by Andreas Kain of BMW and Dr. Carlos Gorbea, Col. U.S. Army, and former student at the Technical University of Munich. Dr. Gorbea's PhD dissertation focused on the application of the methods explored in this case study through his research at BMW.

Supporting Life in Space at NASA
A Case Study for *The Mechanical Design Process*

Introduction

A key need in manned space exploration is a reliable Portable Life Support System (PLSS). These systems provide all of the functions necessary to keep the astronaut alive and comfortable during a spacewalk. The first generation PLSSs were used during the Apollo Program and the image of moon walkers with large back packs is memorable. The second generation, the Shuttle Extravehicular Mobility Unit (EMU) PLSS was designed in the 1970s for use on the Space Shuttle; it has also been used to support International Space Station assembly and maintenance. It was designed for the microgravity of low-earth orbit and had to be small to fit through the hatches on the space shuttle.

It has been over thirty years since a new life support system has been developed and many new technologies have evolved during this period. Further, NASA is now planning visits to Mars and asteroids, as well as a return to the moon. Thus, a project was initiated in 2005 to develop the next generation Advanced PLSS.

The primary challenges in designing the Advanced PLSS are the co-development of maturing technologies and, even more difficult, the integration of these technologies in a small envelope. This is a complex systems integration problem as well as a packaging exercise, as heat management is very challenging in the vacuum of space and minimizing volume is of vital importance in spaceflight applications.

Figure 1: A model of the Advanced PLSS

The Problem: Integrate evolving technologies in a compact envelope

The Method: Engineers are developing a series of prototypes so they can evaluate component and system performance as well as technology readiness in this complex system-of-systems

Advantages: These methods take time, but the final system needs to be highly reliable as astronauts lives depend on them.

Background

Who can forget the images from the moon of Neil Armstrong and Buzz Aldrin exploring the lunar surface with large packs on their backs (Fig. 2)? These "packs" were their Portable Life Support Systems (PLSSs) providing oxygen, regulating pressure, removing carbon dioxide and trace contaminants, controlling humidity and maintaining the temperature of the space suit avionics as well as the astronaut. They also provided radio communications and telemetry to give mission control information on the astronaut's and the PLSS's health.

These Apollo PLSSs, developed in the 1960s were 26 inches (66 cm) high, 18 inches (46 cm) wide, and 10 inches (25 cm) deep. They weighed approximately 84 pounds

Figure 2: Buzz Aldrin with PLSS on his back

(38 kg) on Earth, which, in the Moon's gravity is 14 lb (6.4 kg). Additionally, on top of the PLSS, behind the astronaut's head was a separate unit called the Oxygen Purge System (OPS), an emergency backup system to maintain suit pressure and remove carbon dioxide and heat through a continuous, one-way air flow vented to space. The OPS weighed an additional 41 pounds (19 kg) on Earth or 6.8 lb (3.1 kg) on the Moon. The first four missions (Apollo 11 through 14) were limited to 4 hours. Later, the EVA (Extra-Vehicular Activity) duration was doubled to 8 hours by increasing the size of some of the systems.

The PLSS for the Space Shuttle/Space Station was part of the EMU, the name for the entire space suit. These life support systems, developed in the 1970s are commonly referred to as the Shuttle/ISS EMU PLSS. They weighed 141-160lbs (64 - 72 kg) on Earth and are weightless in the microgravity environment of low-earth orbit. While this system is heavier than the Apollo PLSS, it has more capabilities and utilizes different materials to improve robustness and resist corrosion. Improving the resiliency of the system and preventing corrosion contamination is extremely important on the ISS as the PLSSs are used repeatedly, as opposed to the one time use PLSSs that were discarded on the surface of the moon after the completion of each Apollo spacewalk.

In 2005 NASA began a new, Advanced PLSS design program. The goals of the Advanced PLSS are:
- Simpler, more robust and reliable system design
- Optimized for low-earth orbit and Langrangian point EVA operations.
- Provide flexibility for deep space or lunar missions, and is "Mars forward".
- Generate more sensor data
- Provide EVA capability in more severe situations (e.g. very hot environments)
- Provide additional emergency capabilities (60 minutes, as opposed to 30 minutes in Apollo and Shuttle/ISS PLSSs)
- Weight ≈ 150 lbs

The Advanced PLSS will be a low volume product with less than 20 PLSSs to be built over the life of the product.
The development program is divided into five phases:
- PLSS 1.0 - Breadboard tests to refine subsystems and to be sure that all systems work together. Completed in summer 2011.
- PLSS 2.0 - Fully integrated system assembled in March 2013. To be tested in late 2013.
- PLSS 2.5 - Pre-production prototype to run on Nitrogen
- PLSS 3.0 - System to run on O_2 with human subjects in vacuum chamber
- PLSS 3.1 - Flight test unit for demonstration on ISS

This case study was written during Phase 2.0 testing. From here on, the Advanced PLSS will be referred to simply as the "PLSS".

PLSS – a System of Systems

The PLSS has Oxygen, Ventilation, and Thermal Subsystems that provide the same life support functions as in the Shuttle/ISS EMU PLSS. The schematic for the PLSS is shown in Fig. 3. The important systems and sub-systems are described in this section.

The right side of the diagram is the Space Suit Assembly (SSA) and is outlined in orange. Inside this suit, the astronaut wears a conformal item of clothing known as the Liquid Cooling and Ventilation Garment (LCVG). The LCVG is woven with approximately 300 ft of plastic tubing through which chilled water is circulated to keep the crew member cool. The LCVG also incorporates ventilation ducting that draws the space suit gas back into the PLSS for reconditioning (carbon dioxide and water vapor removal, etc.).

The PLSS is on the left, broken into its three major sub-systems: Oxygen (purple and orange), Ventilation (green) and Thermal Control (blue).

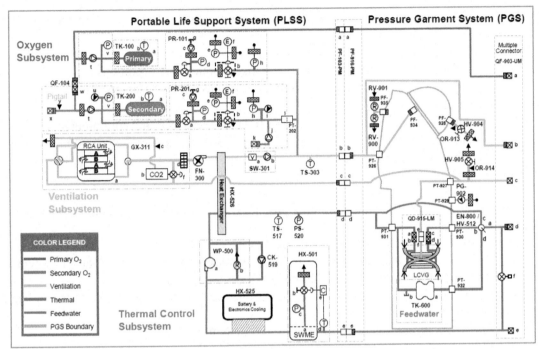

Figure 3: Schematic for the PLSS

The Oxygen Subsystem has both primary (purple) and secondary (orange) high pressure storage tanks and regulators. The primary storage tank holds 200 in^3 (3,277 cc) of O_2 at 3000 psi (320.7 MPa). The pressure regulators not only control the supply of oxygen to the astronaut, but maintain a reasonable pressure inside the suit. One goal of this project was to develop a new Primary Oxygen Regulator (POR).

Where the Shuttle/ISS EMU PLSS had only two pressure set points, the new regulators will have ~4000 possible set points across the full range of 0-8.4 psid. The additional set points provide significant operational flexibility because the suit can be operated at any pressure or slowly transitioned between operating pressures. With respect to suit pressure, there is a tradeoff between physiological habitability and suit mobility. As the suit pressure is reduced the flexibility of the suit increases, making it easier to work; however, there is a lower limit to the habitable pressure range (approximately 2.8 psia, per the current medical literature). The accuracy of the regulation band of the nested set-points of the primary and secondary regulators along with regulator droop (both performance characteristics dependent on the regulator technology) impact the regulator set-points with respect to the physiological limit.

Having the flexibility to change the set-point means that this regulator design can be used regardless of the desired set-pressure in accordance with the mission definition, medical community opinion, or vehicle interfaces. This advanced capability also enables slow transitions between suit pressure set-points that could reduce pre-breathe duration. Prior to an EVA, the astronaut has to complete a "pre-breathe" period during which he or she breathes 100% O_2 to purge dissolved nitrogen from the body's fluids and tissues so that it does not come out of solution when operating inside the space suit at reduced pressure.

This phenomenon is known as decompression sickness or "the bends" and is the same condition to which underwater divers are susceptible. Symptoms of decompression sickness range from pain to paralysis or even death. If symptoms of decompression sickness are observed, treatment involves placing the affected individual under increased pressure. On Earth, such treatment is conducted in a hyperbaric chamber. If such an event were to occur in space, treatment would involve pressurizing the suit as high as possible (8.4 psid) inside the vehicle or habitat (usually at 10.2-14.7 psia); in this manner, having another feasible regulator set-point achieves the in-suit decompression treatment capability.

The Ventilation Subsystem manages carbon dioxide, water vapor, and trace contaminants. Its main component is the Rapid Cycle Amine (RCA) system. The RCA system is a regenerative assembly capable of simultaneously removing carbon dioxide (CO_2) and humidity from the gas stream. Amines are organic compounds that, when exposed to CO_2 or water vapor absorb them, and when exposed to the vacuum of space release them. The RCA has two amine sorbent beds that are alternated between these two modes. During the uptake mode, the sorbent is exposed to the ventilation loop to adsorb CO_2 and water vapor, while during the regeneration mode, the sorbent rejects the adsorbed CO_2 and water vapor to a vacuum source. While one bed is in the uptake mode, the other is in the regeneration mode, thus continuously providing a sorbent bed to remove CO_2 and humidity from the system. This is referred to as a "swingbed" system as the two sorbent beds "swing" alternately between the two modes.

The RCA has the potential of a ~ 65% mass decrease compared to EMU PLSS scrubber which only removed CO_2 necessitating a separate system for humidity control. Additionally, the CO_2 scrubbers on the Apollo and Shuttle/ISS EMU PLSSs utilized chemicals that were consumed as they were used. Thus, CO_2 removal capability could be exhausted during an EVA and was a limiting factor in determining EVA duration. Since the RCA is regenerable real-time during an EVA, CO_2 removal capability will no longer be a consumable or a limiting factor.

One of the most important subsystems is the Thermal Control Subsystem, the PLSS water loop. It must not only maintain the body temperature of the astronaut, but must remove heat generated by fans, pumps, instruments and electronics used throughout the PLSS.

This may seem simple, but there is no possibility of convection with the vacuum of space. Thus, conductive heat transfer between the metal structures within the PLSS and evaporative heat transfer accomplished by the Spacesuit Water Membrane Evaporator (SWME), Fig. 4, are the primary methods of heat rejection. Water circulating through the LCVG and avionics coldplates picks up heat and circulates it through the SWME. The SWME rejects crew and electronics heat by evaporating water through a hydrophobic, porous hollow-fiber membrane. The membrane is formed into channels and the thermal control loop water is passed through them. The pressure difference with the vacuum of space in the vapor channels causes water vapor to pass through the pores in the membrane and then to evaporate. This evaporation cools the remaining water flowing through the channels.

Key requirements on the SWME include removing 810 watts of heat from 91 kg/hr water flow. It must have a volume of 2048 cm^3 (125 in^3) or less, and a mass of 1.59 kg (3.4 lbm) or less while operating in a vacuum of 10-12 torr. The SWME when fully refined is expected to have a 400% improvement of operational life as compared to the Shuttle EMU system. Further, it will dramatically reduce water quality requirements as it is much less sensitive to contamination than the Shuttle/ISS EMU sublimator that provides cooling for the EMU PLSS.

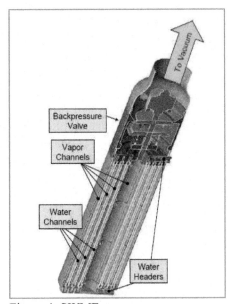

Figure 4: SWME cutaway

Part of the thermal management challenge for the PLSS is cooling the avionics. In PLSS 2.0 they are all contained within a box slightly smaller than a shoebox, which is mounted on a coldplate that interfaces with the water loop, rejecting the heat via SWME. For PLSS 2.5, the team is considering a distributed avionics architecture that will allow the PLSS structure to act to distribute the heat more evenly. This would allow the avionics coldplate to be eliminated from the design, reducing the volume and weight of the system. It is also anticipated that the modular architecture would more readily facilitate component-level testing and software upgrades.

Technology Readiness

One challenge in the development of the PLSS, and true of most evolving systems, is that many of the sub systems are not fully mature. NASA uses a Technology Readiness Level (TRL) scale to communicate maturity (see Appendix A).

At the beginning of the project the RCA, SWME, and O_2 regulators were TRL 1 (basic principles observed) and progressed to TRL 3 (proof of concept) in component-level analyses and functional testing between 2007 and 2010. The components and system progressed to limited TRL 4 (validation in laboratory) in PLSS 1.0.

In PLSS 2.0 these systems will be tested to a higher fidelity TRL 4 because testing will include vacuum, but not all the thermal, vibration, and radiation conditions. PLSS 2.5 will be tested in all relevant environments, but it still will not operate with O_2, so it will not yet be TRL 5 (validation in relevant environment). PLSS 3.0 will bring the testing to TRL 7 (demonstration in operational environment) and PLSS 3.1 (the flight certification unit) will attain TRL 8 (mission qualified). A demonstration on the International Space Station (ISS) will mature all the subsystems and the system of systems to TRL 9 (mission proven).

Thermal packaging

As mentioned in the introduction to the SWME, heat management is a significant PLSS design challenge. Table 1 shows the amount of heat (metabolic rate) generated by the crew in different working conditions. On the first row, 88 watts is human resting heat generated. When running hard a person generates about 1000 watts. For short periods, the crew

Metabolic Rate		CO_2 Injection	H_2O Injection	MGC
(W)	(Btu/hr)	(g/hr)	(g/hr)	(g/hr)
88	300	28	34.7	17.7
117	400	37	44.3	23.6
293	1000	93	82.6	59.1
469	1600	149	88.8	94.5

Table 1: Heat generation data

may generate much more heat than that, so PLSS 2 testing will evaluate more than 2000 watts of heat generation and rejection. Also shown in the Table are the amount of CO_2 and H_2O generated. The word "injection" is used here as the table is taken from the requirements for PLSS 1 where the crew was simulated and the heat, CO_2 and H_2O had to be "injected" into the system.

Additionally, electronics add 80-120 W depending on the mission and battery selection. Batteries are not 100% efficient and the waste heat due the inefficiencies must be removed from them or they will overheat, shortening their lives.

How the heat is managed in the PLSS is highly dependent on the packaging of the components within the envelope. In terrestrial electronics, heat can be dissipated by convection, moving air over the components. Here heat must either be removed by the circulating water in the Thermal Control Subsystem or conducted through the PLSS structure to a place where it can be removed. Thus, the PLSS must be packaged in a minimal space and, at the same time, help manage the heat flow. These two together are referred to as thermal packaging. NASA doesn't have a single method for designing thermal packaging, but has an integrated team of designers, test engineers, and analysts who iterated continually for about 9 months to get to a PLSS 2.0 design that hopefully will change little as the system matures.

- This team used Microsoft Project for project planning and an integrated suite of CAD and analysis tools including:
- Pro/Engineer® (aka Pro/E): a solid modeling CAD package for design work. The solid model was used to generate drawings and code for manufacturing parts, to generate the Thermal Desktop model and as an input to Femap® and Nastran® for stress analysis.
- MacroFlow®: a software tool for rapid and accurate flow and thermal design using for evaluating pressure drop and flow conditions given a particular hardware configuration
- Thermal Desktop®: a design environment for generating thermal models of systems. Thermal Desktop develops the capacitance and conductance network for input to SINDA/FLUINT.
- SINDA/FLUINT®: a finite-difference, lumped parameter (circuit or network analogy) tool for heat transfer design and fluid flow modeling. This system allows detailed thermal modeling and computational fluid dynamics (CFD).
- Femap and Nastran: tools for stress analysis.

Prototype Design and Testing

In order to test the integrated PLSS design and especially to increase the TRL of the new technologies, a series of prototypes are being used.

PLSS 1.0 (Fig. 5), a breadboard was tested from June-September 2011. The testing accumulated 233 hours over 45 days, while executing 119 test points. An additional 164 hours of operational time were accrued

Figure 5: PLSS 1.0

during the test series, bringing the total operational time for PLSS 1.0 testing to 397 hours. As can be seen in Fig. 5, packaging was not considered. The overall PLSS 1.0 test objective was to demonstrate the capability to provide key life support functions. These functions included suit pressure regulation, carbon dioxide (CO_2) and water (H_2O) removal, and thermal control. Specific goals were to:

- Confirm that the components perform in a system-level test as they have performed during component-level testing. The push toward TRL 4.0.
- Identify unexpected system-level interactions
- Operate the PLSS in nominal and off-nominal, steady-state and transient EVA modes with respect to metabolic rate, suit pressure and flow rate, and environmental conditions

PLSS 2.0 was designed and built in 2012 and 2013 and tested in late 2013. It is a fully integrated PLSS design building on lessons learned from PLSS 1.0. It is a complete system as shown in Fig 6, designed to work with nitrogen rather than oxygen. Oxygen will not be used in the entire system until PLSS 3.0, but substantial analysis and sub-component level testing will be conducted throughout the process to verify oxygen compatibility at both the component and system levels. Thus, by the time the team proceeds to PLSS 3.0, a full oxygen compatibility assessment will have been completed.

Figure 6: PLSS 2.0

PLSS 2.0 will be tested with the Space Suit Assembly Simulator (SSAS), a high-fidelity simulator of the space suit volume (Fig. 7). NASA only has a small number of prototype suits and they are continually being used to support various tests, so it would have been impractical for the PLSS team to monopolize one for a year or more. Also, where real suits are very complex adding concerns not important for PLSS testing, the SSAS simplifies the suit-PLSS interfaces for the purpose of development testing.

To build the SSAS the team laser scanned the prototype space suit to generate point clouds of scanned data. Then cleaned up and meshed the point clouds to generate Pro/E CAD files. They then had molds built and plastic injection molded the space suit shown in Fig. 7.

Inside the SSAS they installed an instrumented mannequin. It is wrapped in aluminum tape and then heater tape that simulates the metabolic heat that the crew would generate. The mannequin is wearing a liquid cooling and ventilation garment (LCVG) so that it has the appropriate interfaces to the PLSS water and ventilation loops.

The SSAS and manikin combination well simulates the correct volume, heat generation, gas mixing and other factors of the real space suit and crew member. It is instrumented with equipment to support PLSS 2.0 testing, including interfaces to a simulated vehicle water loop, simulated metabolic gas consumption and suit leakage, interfaces to inject water vapor and carbon dioxide to simulate metabolic products, a displays and controls module, a sound level meter, gas analyzer, etc.

PLSS 2.0 system testing goals are to examine nominal EVA operations; evaluate control algorithms, limit checking, and fault detection requirements; simulate failure modes from FMEA and monitor system response and controls; and evaluate methods of control/user interfaces. Also, as mentioned previously, PLSS 2.0 testing with help the critical new sub-systems reach higher fidelity TRL 4.

Figure 7: The SSAS

Conclusions

The PLSS is a work-in process, an effort to develop a system of systems that includes many new technologies. It uses a series of prototypes to refine the system and the sub systems simultaneously. It is a good example of the use of prototypes and of technology readiness.

Links/References

Packaging Factors for Portable Life Support Subsystems Based. On Apollo and Shuttle Systems, G. A. Thomas, NASA Johnson Space Center, SAE technical Paper 932182. 23rd International Conference on Environmental Systems Colorado Springs, Colorado July 12-15, 1993, http://papers.sae.org/932182/.

Space Suit Portable Life Support System Test Bed (PLSS 1.0) Development and Testing, C. Watts et al. AIAA, 42nd International Conference on Environmental Systems; San Diego, CA; 15-19 Jul. 2012, https://ntrs.nasa.gov/search.jsp?R=20120007410.

Spacesuit Water Membrane Evaporator Development for Lunar Missions, M. R. Vogel, et al, SAE International Conference on Environmental Systems; San Francisco,CA; 2008-402-0313, 2008, https://ntrs.nasa.gov/archive/nasa/casi.ntrs.nasa.gov/20080014235.pdf.

Acknowledgements

Carly Watts, Space Suit & Crew Survival Systems, NASA Johnson Space Flight Center assisted in writing this case study.

Appendix A: NASA Technology Readiness Levels

- **TRL 1 Basic principles observed and reported:** Transition from scientific research to applied research. Essential characteristics and behaviors of systems and architectures. Descriptive tools are mathematical formulations or algorithms.
- **TRL 2 Technology concept and/or application formulated**: Applied research. Theory and scientific principles are focused on specific application area to define the concept. Characteristics of the application are described. Analytical tools are developed for simulation or analysis of the application.
- **TRL 3 Analytical and experimental critical function and/or characteristic proof-of-concept**: Proof of concept validation. Active Research and Development (R&D) is initiated with analytical and laboratory studies. Demonstration of technical feasibility using breadboard or brassboard implementations that are exercised with representative data.
- **TRL 4 Component/subsystem validation in laboratory environment**: Standalone prototyping implementation and test. Integration of technology elements. Experiments with full-scale problems or data sets.
- **TRL 5 System/subsystem/component validation in relevant environment**: Thorough testing of prototyping in representative environment. Basic technology elements integrated with reasonably realistic supporting elements. Prototyping implementations conform to target environment and interfaces.
- **TRL 6 System/subsystem model or prototyping demonstration in a relevant end-to-end environment (ground or space):** Prototyping implementations on full-scale realistic problems. Partially integrated with existing systems. Limited documentation available. Engineering feasibility fully demonstrated in actual system application.
- **TRL 7 System prototyping demonstration in an operational environment (ground or space):** System prototyping demonstration in operational environment. System is at or near scale of the operational system, with most functions available for demonstration and test. Well integrated with collateral and ancillary systems. Limited documentation is available.
- **TRL 8 Actual system is completed and "mission qualified" through test and demonstration in an operational environment (ground or space)**: End of system development. Fully integrated with operational hardware and software systems. Most user documentation, training documentation, and maintenance documentation completed. All functionality tested in simulated and operational scenarios. Verification and Validation (V&V) completed.
- **TRL 9 Actual system "mission proven" through successful mission operations (ground or space):** Fully integrated with operational hardware/software systems. Actual system has been thoroughly demonstrated and tested in its operational environment. All documentation completed. Successful operational experience. Sustaining engineering support in place.

Unsticking A Concept At
MAGICWHEELS
A Case Study for *The Mechanical Design Process*

Introduction

Wheelchairs work well on flat, level surfaces, but on inclines and soft surfaces they can be impossible or even dangerous. Wheelchair users refer to this problem as being in "flat-jail". In 1996 Steve Meginnis set out to resolve this limitation. Steve already had a background in developing products with nearly 40 years of experience designing and developing medical and aircraft products, He holds over 20 patents and was the mechanical engineer who developed the Sonicare ® toothbrush. He considers solving hard mechanical problems a challenge and the development of a wheelchair wheel that could do more than move on a flat, level surface pushed even his capabilities.

Traditional wheelchairs are propelled by pushing on a hand rim which is attached directly to each wheel. The rider pushes, releases, re-grabs the rim, and pushes again. The challenge in negotiating hills and other not-so-nice surfaces is to develop a mechanism that can gear down the tangential force put into the hand rim. This seems simple – cars, bicycles and other devices have had transmissions for over a hundred years. However, when a wheelchair is powered by cyclical arm pushes, especially by people with limited strength and mobility, it is not so simple.

Figure 1: A wheelchair with MAGICWheels going up a hill

MAGICWHEELS® tackled this problem and developed a 2-speed, patented system that has met with wide acceptance. During the development, there were several times when the design got stuck and mechanical problems had to be overcome with clever changes. In this case study, we explore methods to help overcome design sticking points.

The Problem: **MAGIC**WHEELS was running out of time and their product could not be shifted with one hand – a customer requirement.

The Method: Techniques useful to overcome stuck design problems.

Advantages: These methods take work, but they can help un-stick hard design problems.

Traditional wheelchairs are propelled by grasping the hand rim as shown in Figs. 1 and 2. This gives a 1-to-1 gear ratio between the motion of the hand and the rotation of the wheel. In order to gain more torque - needed to go up a hill - the wheelchair user has only one option and that is to apply more force to the hand rim. This puts high stress and strain on the shoulder and arm joints, and muscles, often causing severe repetitive motion injury. Many wheelchair users are limited in the amount of force they can apply due to their physical abilities. Additionally, if part way up a grade and the hand rims are released, the chair will roll back down the hill.

Figure 2: Pushing on the hand rim and details of a MAGICWheels 2-gear wheel

This is not a new problem. The first multi speed wheelchair was patented over fifty years ago and there have been many since, but none have made it into commercial use until **MAGIC**WHEELS. Before exploring solutions to this problem and how Steve and **MAGIC**WHEELS resolved the issue, the problem needs to be better understood and the set of customers' requirements discussed.

Customers Requirements

In developing the customers' requirements, the first question that needs to be answered is "Who are the customers?" (see Chapter 6 in *The Mechanical Design Process*). The primary customers are paraplegics, quadriplegics and other disabled persons who use wheelchairs to get around. Beyond these primary customers, insurance companies, Medicare/Medicaid and the Veterans Administration are important as they pay for most mobility aids and accessories.

What Steve wanted to develop was the ability to operate a wheelchair on hills and other uneven terrain. Ideally a wheelchair should be able do the following sequence of events:

- Cruise on a level surface as with a traditional wheelchair,
- Easily shift into a lower gear when approaching a hill,

- Negotiate the hill with no more shoulder load than when on flat ground,
- Hold its position when the hand rim is released going up the hill,
- Easily shift back into cruise when cresting the hill,
- Assist in braking when going downhill, so the chair can't run away.

These translate into the following engineering requirements:
- 1:1 gear ratio through hand rim for normal use

Any geared wheel system must behave like a traditional wheelchair when on a flat surface which is at least 95% of the time. It must be virtually invisible (functionally) to the user until shifted to the lower gear when needed.

- Take load off arms and shoulders for climbing hills, going up ramps, etc
This requires sufficient gearing to climb 5% grade with same or lower force as on a level ground in 1:1 gear. A maximum grade of 5% is the American with Disabilities Act (ADA) requirement. (**MAGIC**WHEELS advertises 10% and actually tests on 15%).

- Easy, one-handed shifting between gears
Shifting should be with similar motions as is propulsion to ensure rapid transition between gears. Since the two wheels are independent, shifting must be with one hand for each wheel.

- No roll back – must automatically hold when climbing hills
When climbing hills it is important that the wheelchair hold its position between power thrusts on the hand rim and not require any other action to hold position if the user pauses.

- Minimal "windage" – low or no increase in drag due to gearing when it is not in use.
When in direct drive (1:1) there must be no drag added by the system, and conversely, when in low gear no energy should be wasted on the 1:1 ratio.

- Downhill control
If the product is to go up-hill, it must also help in going downhill. On many hills, riders can not create enough hand rim friction force on standard wheels to control the wheelchair.

- Minimal additional weight
The capability for going up-hill with lower gears can not offset a major increase in additional weight. Target is <10lbs additional weight, total for both wheels.

- Fits on most common wheelchairs
There are a wide variety of wheels chairs manufactured and there are no standards for how the wheels are connected to them. Thus the resulting product must be universal or have fittings to enable it to fit most chairs.

- Payment approved by insurance, Medicare/Medicaid and VA
Virtually all mobility aids are paid for by insurance companies, Medicare/Medicaid or the Veterans Administration. Thus, it is imperative that the resulting product be

something they approve and will pay for. Typically Medicare pays approx. 80% of the purchase cost and the occupant is responsible for the remaining 20% or this may be covered by secondary insurance.

This is a difficult set of requirements and it took **MAGIC**WHEELS over 10 years to resolve them and bring a product to market. Videos, referenced at the end of this Case Study, show how well **MAGIC**WHEELS met them. The rest of this case study focuses solely on those requirements that Steve and his colleagues struggled with.

Meeting the Requirements

There have been many attempts to meet the customers' requirements over the years. A patent search has turned up over 20 different designs. These concepts range from complex lever systems, to planetary gear sets much as those used in 3-speed bicycle hubs and in the Ford Model-T transmission (a two speed transmission much as needed here). Steve and his colleagues at **MAGIC**-WHEELS studied these patents and realized that none of them could meet the requirements. For most, the interface between the user and wheelchair was complex. For others, the inability to control roll-back, lack of hill-holding and windage were critical.

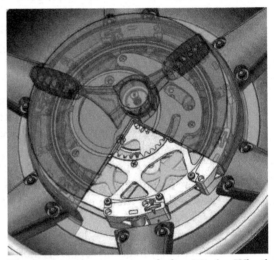

Figure 3: A cutaway of the Magic Wheel transmission

What became the **MAGIC**WHEELS company began life with research efforts at the University of Washington in 1993-94. The company, **MAGIC**WHEELS, was founded in 1996 to refine these 2-speed ideas and bring them to market. Government grants to fund the development effort were applied for and initially rejected. Finally after two years, grants through the NIH and other government agencies were secured. To date, about $2M in development money has been expended.

The key to **MAGIC**WHEELS is the use of a hypocycloidal gear train with a ring gear, a spur gear and a hold gear (Fig. 3). The ring gear is directly affixed to the wheel. The spur and hold gears move in and out of engagement in a plane with the ring gear depending on the position of the shifter (Fig. 5). The spur gear is engaged for the low-gear configuration (push the shift handle forward, clockwise in Fig. 4) and orbits the inside of the ring gear but does not rotate on its own axis; it is fixed to the non-rotating orbiting plate behind it. The hold gear is engaged for the 1:1 configuration (pull the shift handle back) but just locks the wheel and gears into a solid rotation. The mechanizm for this is quite complex and beyond this article. It can be seen in the cutaway in Fig. 3, but is best appreciated by viewing the on-line videos referenced at the end of the article.

The resulting system has over 300 parts (Fig. 4). Before concluding that this is overly complex for the function it serves note that before **MAGIC**WHEELS, no system could achieve the requirements. Further, before **MAGIC**WHEELS, Medicare had only two code categories for Wheelchairs, Manual and Powered. There was no category to cover multiple speed manual wheelchairs. It took **MAGIC**WHEELS over two years of working with Medicare to get a code established for this classification (created in Jan 2008).

During the development of this system, once committed to the hypocycloid, progress got stuck in that mechanism did not shift well enough to bring to market. It was necessary to move the hand rim while moving the shifting lever back & forth to get the gears to mesh. This motion required two hands for each wheel violating an important design requirement. The design team worked on this problem for six months. As a funding deadline approached they knew they had to resolve the issues. They tried many things, each helped a little. For example, they changed the shape of the gear teeth to allow them to not only engage and disengage more easily, but to make shifting so that the new gearing is made before the old one is release to maximize safety. As time was running out, Steve noticed that the hand rim only needed to move in one direction to initiate shifting - in the opposite direction of the natural motion shown in Fig. 5. It was 2:00AM one morning that he realized that reversing the system would alleviate most of the problems.

Figure 4: An exploded view of the MAGIC-Wheels transmission

Figure 5: The shifting action needed

So they reversed the parts (left to right) and put the transmission together backwards. With this change, the hand rim and shift handle move in the same direction during shifting and the friction on the gear box from moving the shift handles also moves the hand rim slightly in the same direction providing smooth single handed shifting. This "reversing of the parts" resolved the problem and afforded one hand powering and shifting on each wheel. It allows a wheelchair user to approach a hill in high gear, coast up a little and then hold the shift levers as the wheelchair stops and coasts backward slightly thus shifting into the low gear which stops the backward motion as it is automatically in hill-holding mode. The entire shifting motion was now close to the ideal painted by the requirements.

A young colleague asked Steve how he thought of reversing the parts and he told him that you must try everything. He knew there had to be a solution and if you look long enough you will find one. He spent many months of mental and physical trial and error to solve the problem.

Steve has over 40 years of design experience. He has learned many tricks for un-sticking his thinking. Common methods for un-sticking problems are recapped and related to Steve's experience in the next section.

Un-sticking the Thinking

Typically, when facing a dilemma during design, the situation is something like:" If I improve X, I will adversely affect Y, How can I improve them both?" This is often referred to as a "trade-off" "conflict" or "contradiction". Two methods for articulating and studying contradictions are: a) using patching guidelines, and b) use TRIZ. Each of these can help un-stick design problems caused by contradictions.

"Patching" is a term   widely used in software development for fixing contradictions. There are eight patching guidelines that work well with mechanical devices (see Section 9.3.5 in *The Mechanical Design Process*). They are:

1. *Combining:* Make one component serve multiple functions or replace multiple components. Combining will be strongly encouraged when the product is evaluated for its ease of assembly (Section 12.5).
2. *Decomposing*: Break a component into multiple components or assemblies. As new components or assemblies are developed through decomposition, it is always worthwhile to review constraints, configurations, and connections for each one. Because the identification of a new component or assembly establishes a new need, it is even worthwhile to consider returning to the beginning of the design process with it and considering new requirements and functions.
3. *Magnifying/Minifying*: Make a component or some feature of it bigger/smaller relative to adjacent items. Exaggerating the size or number of a feature will often increase one's understanding of it.   Make one dimension very short or very long. Think about what will happen if it goes to zero or infinity. Try this with multiple dimensions. Sometimes eliminating, streamlining, or condensing a feature will improve the design.

4. *Rearranging*: Reconfigure the components or their features. This often leads to new ideas, because the reconfigured shapes force rethinking of how the component fulfills the functions. It may be helpful to rearrange the order of the functions in the functional flow. Take the current order of things and switch them around. Put what is on top, on the bottom; or what is first, last.
5. *Reversing*: Transposing or changing the view of the component or feature; it is a subset of rearranging. Try taking what is the inside of something and making it the outside or vice versa. Or try switching left for right.
6. *Substituting*: Identify other concepts, components, or features that will work in place of the current idea. Care must be taken because new ideas sometimes carry with them new functions. Sometimes the best approach here is to revert to conceptual design techniques in order to aid in the development of new ideas.
7. *Stiffening*: Make something that is rigid, flexible or something that is flexible, rigid.
8. *Reshaping*: Make something that is first thought of as straight, curved. Think of it as cooked spaghetti that can be in any form it wants to be and then hardened in that position. Do this with planar objects or surfaces.

Relating these to Steve's situation at **MAGIC**WHEELS, what solved his problem was #5 - Reversing. However, before that he had reshaped the gear teeth (#8).

In fact these types of patches are so common, that a Russian patent inspector realized that they were the essence of most good ideas. After he reviewed many thousands of patents, he developed a very complete list of, what he called, the 40 Inventive Principles. He found that these 40 inventive principles underlie all patents. These are proposed "solution pathways" or methods of dealing with or eliminating engineering contradictions between parameters.

The Inventive Principles are part of TRIZ (pronounced "trees") the acronym for the Russian phrase "The Theory of Inventive Machines." This method developed by Genrikh (or Henry) Altshuller, a, mechanical engineer, inventor and Soviet Navy patent investigator. After WWII Altshuller was tasked by the Russian government to study world-wide patents to look for strategic technologies the Soviet Union should know about. He noticed that some of the same principles were used over-and-over again by totally different industries, often separated by many years, to solve similar problems.

From his findings Altshuller began to develop an extensive "knowledge-base" which includes numerous physical, chemical, and geometric effects along with many engineering principles, phenomena and patterns of evolution. Altshuller wrote a letter to Stalin describing his new approach to improve the Rail System along with products the USSR produces. The communist system at the time didn't value creative/free thinking. His ideas were scorned as insulting, individualistic and elitist, and as a result of this letter, he was imprisoned in 1948 for these capitalist and "insulting" ideas. He was not released until 1954, after Stalin's death. After the 1950s, he published numerous books, technical articles, and taught TRIZ to thousands of students in the former Soviet Union.

Altshuller's initial research in the late 1940's was conducted on 400,000 patents. Today the patent data-base has been extended to include over 2.5 million patents. This data has led to many TRIZ methods. Only part of the most basic one will be described here. This method makes use of Contradictions and Inventive Principals. The links at the end of the paper give a complete list and description of the principles, a super-set of the eight patching guidelines.

The first step in using TRIZ is to discover the contradiction. In Steve's case it was evident, when the gears are shifted; the motion is in the wrong direction. In many cases the contradiction is not so evident and so there are methods within TRIZ to discover them (A method to find contradictions is presented in Section 7.6 in *The Mechanical Design Process*)

The second step is to use contradictions as an index to the 40 principles to discover which may apply. This indexing method is not covered here. Rather a couple of the Inventive Principles found through using it will be discussed. These inventive principles are:

Principle 2. Inversion
   a. Instead of an action dictated by the specifications of the problem, implement an opposite action
   b. Make a moving part of the object or the outside environment immovable and the non-moving part moveable
   c. Turn the object upside down
   Example:
   1. Abrasively clean parts by vibrating the parts instead of the abrasive

This is very much like patching guideline 5, Reversing. As said earlier, the guidelines are a subset of the TRIZ Principles. Clearly this principle can help focus thinking in the direction that ultimately solved Steve's **MAGIC**WHEELS problem.

Other potential solutions suggested by the indexing the Inventive Principles are:

Principle 3. Mechanical vibration
   a. Set an object into oscillation
   b. If oscillation exists, increase its frequency, even as far as ultrasonic
   c. Use the frequency of resonance
   d. Instead of mechanical vibrations, use piezovibrators
   e. Use ultrasonic vibrations in conjunction with an electromagnetic field

Examples:

1. To remove a cast from the body without skin injury, a conventional hand saw was replaced with a vibrating knife
2. Vibrate a casting mold while it is being filled to improve flow and structural properties

Principle 14. Spheroidality

    a. Replace linear parts or flat surfaces with curved ones, cubical shapes with spherical shapes

    b. Use rollers, balls, spirals

    c. Replace a linear motion with rotating movement, utilize a centrifugal force

    Example:

       1. Computer mouse utilizes ball construction to transfer linear two axis motion into vector motion

Clearly each of these Principles suggests other solutions. Principle 3 suggests jiggling the hand rim, the action that helped Steve discover his ultimate solution. And, Principle 14 suggests changing the shapes of surfaces and interfaces – Steve changed from a linear shifter to a rotary one in going from the first generation prototype to the 2nd and he changed the gear tooth shapes to improve shifting. There may be many other ideas in the 40 Principles.

Conclusion

Although Steve did not formally use the eight patching methods or TRIZ, his years of experience have instilled many of these guidelines and Principles in his thinking. As he struggled to un-stick the problem, he combined, reversed, and changed the shapes of things until he finally hit on the solution. Formalizing contradictions and using the guidelines and Principles can shorten the time to a solution and the quality of it. The final **MAGIC**WHEELS product is creating a new market and is receiving strong endorsement from wheelchair users who are escaping from flat-jail.

Resources

Videos showing **MAGIC**WHEELS structure and use can be found on YouTube:

*   http://www.youtube.com/watch?v=fpqmmWjxuEE
*   https://www.youtube.com/watch?v=vJlhfaQlI3Y
*   https://www.youtube.com/watch?v=pyBx60rlaVc

TRIZ 40 Inventive principles are at

*   *The Mechanical Design Process* web site <www.mechdesignprocess.com>
*   www.triz-journal.com. The TRIZ Journal is a good source for all things TRIZ.

Acknowledgements

Steve Meginniss, founder and CTO of **MAGIC**WHEELS, Inc. in Seattle, Washington assisted in the development of this case study..

Autodesk® Inventor® sponsored the development of this case study.

Redesigning the Ceiling Fan at the Florida Solar Energy Center

A Case Study for *The Mechanical Design Process*

Introduction

Ceiling fans are an inexpensive way to cool a room making it feel 2-4°F cooler just by moving the air. The earliest electric powered fans appeared in the 1880s and the basic design of an electric motor with a set of tilted, flat blades has remained virtually unchanged until quite recently when Danny Parker set out to create a ceiling fan design to increase air moving efficiency while reducing energy consumption.

The resulting fan, sold commercially as the Gossamer Wind, consumes about half the energy as previous fans while delivering more air movement. Since its introduction to the market in 2001 more than 2 million Gossamer Winds have been sold.

Figure 1: The Gossamer WInd Ceiling Fan

This case study is about the invention and evolution of this new technology, basically bringing an old technology to a mature product.

The Problem: Design a more efficient ceiling fan. Current ceiling fans, while providing inexpensive cooling, are not very efficient.

The Method: The engineers needed to: discover a new idea, refine the idea to practice, develop and test prototypes and refine the prototype to practice.

Advantages: The companies involved were small and few formal design methods were used. This informality worked because so few people were involved and the organization relatively uncomplicated. This is a good example for the use of informal methods when the conditions are right.

111

Background

The first ceiling fans were used by the Egyptians and Romans. They were powered by their slaves. In 1882, Philip Diehl, who had previously engineered the first electric motor used on sewing machines, adapted that same technology to a ceiling fan. Since then, typical ceiling fans have been essentially unchanged, consisting of a low rpm motor with three - five flat blades attached as

Figure 2: A typical traditional ceiling fan

shown in Fig. 2. Many also have a light or lights in the middle. Common ceiling fans move 60 - 70 CFM of air per watt of electricity used. They usually have multiple speeds ranging from 50 – 200 rpm. Ceiling fans are a very mature product with the only differences throughout their history being their style, number of blades, materials and finish, and lighting options.

Danny Parker, a principal research scientist at the Florida Solar Energy Center (FSEC), part of University of Central Florida, was sitting on his porch one day in the early 1990s with his father-in-law, a WWII fighter pilot. They were sitting under a ceiling fan when his father-in-law noted that the blades were flat and he had never flown an airplane with a flat bladed propeller. Parker is an expert on heating and cooling, but had no aerodynamics expertise. His father-in-law's comment started him wondering why ceiling fans had flat blades and every other propeller was airfoil shaped and twisted.

Parker did some quick research and learned that flat "paddle blades" are easy to manufacture and while simple, they are not very efficient. Further, he learned that airplanes, ships and other propeller driven vehicles <u>all</u> have twisted, airfoil shaped blades. This has been the case since Wilbur Wright developed the propeller for the Wright Flyer in 1903.

His reading told him that he could not simply use an airplane propeller for a ceiling fan. Airplane or boat propellers turn at 2000 – 3000 rpm, or faster. Ceiling fans rotate at a low rpm for safety reasons. Namely, the Underwriters Laboratories (UL), a product safety testing organization, publishes standards for all types of electrical and other products. Ceiling fans are covered in UL Standard 507 which limits the rotational rate to 50 – 200 rpm to avoid severe injuries in case someone comes in contact with a blade. Thus, the design of an efficient ceiling fan blade is more difficult than just applying airplane propeller theory to problem.

None-the-less, based on his initial research, he was convinced that with the right blades, fans could be much more efficient and he convinced others at FSEC that this was an idea that could help energy conservation in Florida.

Even though Parker did much reading and study on propeller technology and history, he found little to help him at the low fan speeds. What he did find was that, in the 1970s a company, AeroVironment, had designed the Gossamer Condor, the first human powered airplane to achieve controlled and sustained flight (1977) and the Gossamer Albatross, (Fig. 3), the human-powered aircraft that was peddled across the English Channel in 1979. The

Figure 3: The Gossamer Albatross

propellers on these craft rotated more slowly than those on conventional propeller airplanes, at about 110 rpm, similar to the need for the fan. Thus, he got in touch with the designer of the Gossamer airplane propellers, Bart Hibbs, who was intrigued by the potential.

Since there were no customers demanding more efficient ceiling fans or any source for research funding, this project was undertaken internally because Parker and Hibbs thought it had potential and they both worked at organizations that would allow them to tinker with new ideas with internal funding. The two organizations, the Florida Solar Energy Center and AeroVironment entered into a business agreement. Basically, they would jointly explore, at a low level, the potential of developing a ceiling fan blade using airfoil theory and see if this might lead to more efficiency. If it did, and led to a product, they would split any royalties along an agreed percentage.

Conceptual design

The first question to be answered was: is there a potential to significantly increase the efficiency of a ceiling fan by using propellers rather than paddles? In early 1997 Hibbs explored this question using proprietary computational fluid dynamics (CFD) models specifically designed for very low speed propeller analysis. Hibbs developed eight fan blade designs using different airfoils and blade shapes and twist. The leading candidate for the ceiling fan design is shown in Fig. 4. The nearly vertical section is near the root of the blade with the chord (leading edge to trailing edge distance) getting smaller and more horizontal toward the tip. Comparing this to a flat, paddle type blade of constant chord (Fig 2), the difference is dramatic. What is even more important, Hibbs' models showed that this blade could move twice the air for the same power, a 100% increase in efficiency.

Prototype Development and Testing

The concepts generated by Hibbs at AeroVironment were evaluated by FSEC and one chosen for testing based on promising theoretical performance. Research engineer Jeff Sonne, at FSEC, was an experienced wood-worker and carefully crafted a set of balsa wood blades while Parker developed a ceiling fan test facility. The

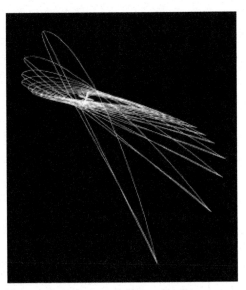

Figure 4: Conceptual design for the Windward

facility allowed measuring the total air flow, the airflow at different radii and the sound level. The test rig was made so that the tip angle could be adjusted as the CFD simulations showed optimal airflow at anywhere between 4 and 6.5 degrees at the tips for a 52" diameter fan (a typical ceiling fan diameter). The test blades had a chord of approximately 6.5" wide at the root tapering to just 2.5" at the rounded tip. The blades were 20.5" long with an estimated surface area of 93 square inches. Unlike the paddle bladed ceiling fans, the propellers for the new design were highly twisted and tapered airfoils.

Testing was done by comparing the new design, called the CF-1 ("Ceiling Fan #1"), to three conventional ceiling fans:

- *Emerson "Northwind" CF705, a low cost model* - $70, representing a very large part of the current ceiling fan market. It consists of a fan with four flat blades and an inexpensive 50 watt shaded pole induction motor. The flat blades had a nominal angle of 12.5 degree, a chord of 5" at the blade root and 5.5" at the tip. They were 20 inches long, made of painted wood and each had a measured weight of 329 grams with mounting hardware.

- Emerson *"Premium" CF4852, a* more expensive (>$400) 5-bladed, 50" diameter fan which represented the upper end market designed to move more air. Advertising claimed 5595 cfm at 100 watts resulting in 56 cfm/watt.

- *Hunter "Summer Breeze"* rated as one of the best ceiling fans in terms of air moving efficiency by <u>Consumer Reports</u>. The flat ceiling fan blades were similar to the Emerson models.

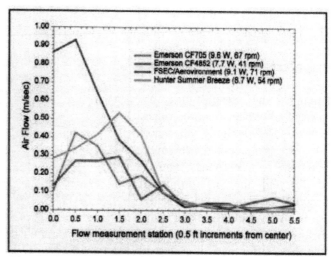

Figure 5: Low speed air flow results

The prototype CF-1 fan used the same motor as the *Emerson CF705* fan with the only difference being the new blades. Testing was primarily done at two speeds, low and high. Low speed varied by the fan model and motor configuration between 40 and 70 rpm. The measured air flow in meters/second for the four different fans at low speed operation over the six foot region comprising the measurement area is shown in Fig. 5. This data is shown in summary form in Table 1. The FSEC/*AeroVironment* fan exhibited clearly superior performance to the paddle blade fans. The Hunter fan provides the most competitive comparable performance. Since the Emerson CF705 and the CF-1 use the same motor it is possible to see the impact of the improved blades alone by comparing their performance results.

Table 1: Comparative fan performance and efficiency at high speed

	Emerson CF705	Emerson CF4852	Hunter Summer Breeze	FSEC/ Aerovironment CF-1
CFM	3110	6057	5339	6471
Watts	50.2	93.1	74.8	49.6
CFM/Watt	61.9	65.1	71.4	130.5

The improvement due to the advanced fan blade shown in Table 1 was particularly dramatic with an efficiency (cfm/watt) improvement of nearly 100% over the CF705, the fan with exactly the same motor.

A typical airflow distribution is as shown in Fig. 6. This clearly illustrates the increased volume of air moved along with a slightly greater radius of perceptible air flow below the disk of motion.

Finally, the test fan was noticeably quieter than the other fans. The Hunter "Summer Breeze" produced 5.7db over ambient background noise (57db) while the CF-1 produced a sound pressure level of only 0.3db over ambient. This 5.4db difference translates into an intensity difference of a factor of 3.5. Since inefficiency is energy that is translated into noise, this is not surprising.

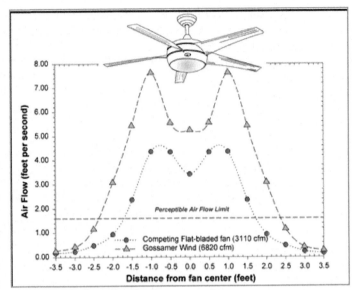

Figure 6: Air flow distribution comparison

These results confirmed what Parker had suspected, there was much efficiency to be gained in the design of the blades on a ceiling fan.

The Patent

In 1998 Parker and his partners at AeroVironment applied for a patent titled "High Efficiency Ceiling Fan". This was awarded in 2000 as patent number 6,039,541. The front page of the patent is shown in Fig. 7.

The 16 claims are all built around the first one (Fig. 8) as is typical in most patents. There are three major ceiling fan blade features claimed: blade twist, blade thickness taper, and chord taper.

United States Patent [19]

Parker et al.

[11] **Patent Number:** 6,039,541

[45] **Date of Patent:** Mar. 21, 2000

US006039541A

[54] **HIGH EFFICIENCY CEILING FAN**

[75] Inventors: **Danny S. Parker**, Cocoa Beach, Fla.; **Guan Hua Su**; **Bart D. Hibbs**, both of Monrovia, Calif.

[73] Assignee: **University of Central Florida**, Orlando, Fla.

[21] Appl. No.: **09/056,428**

[22] Filed: **Apr. 7, 1998**

[51] Int. Cl.[7] F04D 29/38
[52] U.S. Cl. .... 416/223 R; 416/5
[58] Field of Search .... 416/5, 223 R

[56] **References Cited**

U.S. PATENT DOCUMENTS

D. 355,027	1/1995	Young .	
D. 382,636	8/1997	Yang .	
1,506,937	9/1924	Miller	416/223 R X
1,903,823	4/1933	Lougheed	416/223 R X
1,942,688	1/1934	Davis	416/223 R X
2,283,956	5/1942	Smith	416/223 R
2,345,047	3/1944	Houghton	416/223 R
2,450,440	10/1948	Mills	416/223 R
2,682,925	7/1954	Wosika	416/226 R X
4,197,057	4/1980	Hayashi	416/242
4,325,675	4/1982	Gallot et al.	416/223 R
4,411,598	10/1983	Okada	416/223 R
4,416,434	11/1983	Thibert et al.	416/223 R X
4,730,985	3/1988	Rothman et al.	416/223 R X
4,782,213	11/1988	Teal .	
4,844,698	7/1989	Gornstein et al.	416/223 R
4,892,460	1/1990	Volk .	
4,974,633	12/1990	Hickey	137/561 R
5,033,113	7/1991	Wang .	
5,114,313	5/1992	Vorus .	
5,244,349	9/1993	Wang	416/5 X
5,253,979	10/1993	Fradenburgh .	
5,951,162	9/1999	Westman et al.	416/223 R X

FOREIGN PATENT DOCUMENTS

19987	1/1930	Australia	416/223 R
1050902	1/1954	France	416/223 R
676406	7/1952	United Kingdom	416/223 R
925931	5/1963	United Kingdom	416/223 R
92/07192	4/1992	WIPO	416/223 R

Primary Examiner—John E. Ryznic
Attorney, Agent, or Firm—Brian S. Steinberger; Law Offices of Brian S. Steinberger

[57] **ABSTRACT**

Ceiling fan blades for low speed fan operation. The blades have a positive twist at the root motor portion of the blade and a slightly twisted rounded tip. The chord of the blades taper down from the root to the rounded tip, and have a tapered airfoil from the aft forward aft edge to the trailing edge. The airfoil has a combination of a rounded leading edge with sharp trailing edge, and a square leading edge and rounded trailing edge. The blades can be twenty inches in length and twenty-six inches in length, and be used in ceiling fans having two, three, four or more blades in a ceiling mount. The ceiling fan blades are optimized to operate in ceiling fans running at low speed ranges of approximately 50 to approximately 200 revolutions per minute(rpm) with an enhanced axial airflow which provide substantial energy savings and increased air flow over conventional flat planar ceiling fan blades.

16 Claims, 12 Drawing Sheets

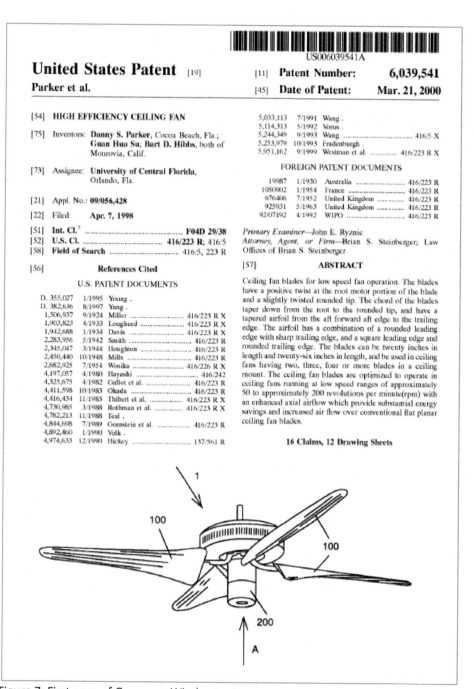

Figure 7: First page of Gossamer Wind patent

Product development

Neither FSEC nor AeroVironment were in the business of producing consumer products. Thus, they needed a third partner to manufacture and market fans based on the new design. They wrote a prospectus which they sent to companies who produced ceiling fans. After negotiations with a number of manufacturers, a business deal was struck with King of Fans, one of the leading producers of home comfort products worldwide. While their name is not well known, they produce over 4 million units a year, selling fans through Home Depot, Lowes and others under a wide variety of names, most notably; Hampton Bay. King of Fans is a family owned Taiwan company with a small group of engineers in Taichung.

I claim:

1. A ceiling fan blade for low speed operation at approximately 50 to approximately 200 revolutions per minute(rpm)
15 for use in overhead ceiling fan systems, the blade comprising:

a positive twist adjacent to a rotor end of the blade, so that blade pitch increases from a tip end of the blade to the rotor end of the blade;

20 an airfoil having a thicker portion at the rotor end tapering down to a thinner portion at the tip end, for providing high lift and low drag; and

a blade chord having a wider portion from the rotor end
25 tapering down to a narrower portion at the tip end, wherein the blade is operated in a ceiling fan running at low speed ranges of approximately 50 to approximately 200 revolutions per minute(rpm) for use with ceiling fan systems, and provides substantial energy
30 savings and increased air flow over conventional ceiling blades.

Figure 8: First claim of patent

These engineers developed a set of specifications needed to turn the CF-1 into a product. They were:

- A significant increase in efficiency. (The test fan produced nearly 100% improvement but they knew this would be lowered in meeting other specifications)
- Meet Energy Star airflow efficiency requirement of 75 cfm/watt at high speed
- Must operate effectively from 50 rpm to 200 rpm
- Must work in pulling air up and pushing it down (forward and reverse). The need to rotate in both directions required additional modification to the fan-blade design. The test airfoil was designed to only move air in one direction.
- No sharp edges (UL requires that no edge of a fan blade be thinner than 3.175 mm (1/8 inch) diameter. The test blades had sharp trailing edges.
- Quietness of operation – at least as good as or better than current fans.
- Provide light
  o At least 1500 lumens
  o High efficiency
  o Long life bulb(s)
  o Easy to replace
- All features should speak of quality
- All features should speak of efficiency

Further, they saw this product as an opportunity to add a remote control device that they had been developing for a number of years. This device should meet the following specifications:

- At least 3 speeds
- Direction control
- Light control

In addition to the specifications, a key consideration of the product was its overall appearance. King of Fans felt that the design that was optimized for airflow and airflow efficiency was too radical in form to generate the level of customer appeal (i.e. sales) necessary at mass retail, so they sacrificed some performance in exchange for a more mainstream appearance. They worked over an eight month period to find the right balance of form and function, a product that could meet the specifications and was visually appealing. The highlights of the resulting product and the effort are described below.

The Blades

The blades on virtually all previous ceiling fans were cut from flat material. Those on the test model were laboriously hand built over a couple of month period. Due to the complexity of the shape, the only reasonable option for production was injection molding. ABS was selected as it is inexpensive and easy to injection mold.

To meet the requirements to produce flow in both directions and have no sharp edges, the trailing edge of the blade had to be rounded. Further, the area near the root had to be flattened (less angle of incidence) to keep the mold from being overly thick, increasing the manufacturing costs. These compromises to meet safety concerns and the manufacturing reality reduced the increase in efficiency to 40%. The production fans still move 6,779 cfm with 64 watts on high speed resulting in 105 cfm/watt. This is still a dramatic increase over existing fans, and the Gossamer Wind exceeds the Energy Star airflow efficiency requirement of 75 cfm/watt at high speed by 40%.

Figure 9: The Gossamer Wind with light activated

Energy Efficient Light Kit

Since this product was to be marketed as eco-friendly and had such a high increase in air moving capability, the lighting efficiency also needed to be addressed.

119

Conventional fans use either a 100W linear halogen lamp or three or more 40W incandescent bulbs. These 100-120 watt lights more than doubled the energy usage of the fan itself (64 watts).

The bulb chosen is to address this problem was a circular compact fluorescent bulb. This uses 20 watts and provides about the same lumens as the halogen or 3- 40 watt incandescent bulbs (1450 lumens), although with a somewhat softer illumination (Fig. 9).

Equally important is the life expectancy: the compact fluorescent lamp has a MTBF of 12,000 hours versus 2,000 hours for the 100 W halogen lamp. Thus, the CFL lamp will likely not need changing over the life of the fan.

Conclusions

The resulting product, manufactured by Hampton Bay, has been commercially available throughout the United States since 2001. The new blade design referred to as Gossamer Wind provides 40 percent higher airflows without increasing energy use. It is one best-selling ceiling fan designs in the market place and, in its thirteenth year, one of the longest offered. It was also one of the first ceiling fans to be Energy Star listed. Since the introduction of the Gossamer Wind other manufacturers have developed similar fans and there is on-going litigation regarding patent infringement within the ceiling fan market.

Links

- This case study borrowed heavily from *Development of an Enhanced Ceiling Fan: An Engineering Design Case Study Highlighting Health, Safety and the Environment* a case study developed by Marc Rosen, Professor at the University of Ontario, Institute of Technology, Jan 2009. http://safetymanagementeducation.com/wp-content/uploads/2015/06/ Minerva_case_study_Enhanced_Ceiling_Fan_090111.pdf
- US Patent 6039541, *High Efficiency Ceiling Fan*, Inventors D. S. Parker, G.H. Su, B. D. Hibbs, 2000
- Development of a high efficiency Ceiling Fan "The Gossamer Wind", D.S. Parker, M. P. Callahan, J. K. Sonne and G.H. Su, Florida Solar Energy Center (FSEC), FSECV-CR-1059-99, 1999, http://www.fsec.ucf.edu/en/publications/html/FSEC-cr-1059-99/index.htm.
- Performance and Applications of Gossamer Wind™ Solar Powered Ceiling Fans, M. Lubliner, J. Douglass, D. S. Parker, D Chasar, FSEC-PF-411-04, 2004 http://www.fsec.ucf.edu/en/publications/html/FSEC-PF-411-04/index.htm.
- Gossamer Wind web site, http://www.gossamerwind.com/content/gossamer-windward-iii-white-0
- Florida Solar Energy Center, http://www.fsec.ucf.edu/en/
- AirVironement, http://www.avinc.com/

- Energy Star Program Requirements for Residential Ceiling Fans: Partner Commitments, http://www.energystar.gov/ia/partners/prod_development/revisions/ downloads/ceil_fans/Final_Ceiling_Fans_Prog_Req_Version_2.0.pdf

Acknowledgements

Danny Parker, FSEC; Professor Marc Rosen, Ontario Institute of Technology, and Tien Lowe, King of Fans assisted in writing this case study.

Idea to Product in One Day
for Pedal Petals

A Case Study for *The Mechanical Design Process*

At breakfast, Sally had an idea for a product. By dinner time she shipped the first Pedal Petal clip-on flower to a customer. This amazingly short time-to-market has only been possible the last couple of years with the advent of Additive Manufacturing (AM). Often called 3D printing and formerly called rapid prototyping, additive manufacturing is changing how products are designed and made.

Pedal Petals are relatively simple products, a clip and a flower, but ten years ago it would have been impossible to get a sample of the clip in under a month and even five years ago, in under a week. Yet Sally was in production in less than one day.

In this case study, we detail how Sally, and her husband Patrick, did this rapid turn and how new technologies are changing the product delivery process.

The Problem: Develop and deliver a product between breakfast and dinner.

The Method: Technologies such as solid modeling, additive manufacturing, and on-line market places make product realization fast.

Advantages: The methods used allowed the designers to go from idea to market in a day, and continue to optimize after product launch.

Figure 1: Pedal Petal flowers on handlebars

Introduction

Prior to the Industrial Revolution, products were made one at a time in a "cottage industry", each made individually and no two exactly alike. Cottage Industry products could be individualized, meeting the exact needs of each customer. Product improvements happened slowly as one manufacturer tried something; was successful, and then others followed.

The Industrial Revolution led to consolidating manufacturing operations in factories leading to efficiency of scale. Costs came way down, but this industrialization required products to be standardized, designed for mass production so every item was the same as every other item. Mass production forced products to be reduced to common denominators so a majority of the customers would be satisfied, but little could be customized. Every example of a specific brand of cell phone looks the same and works the same. The only variations are cosmetic (e.g. the color of the case) and which apps are installed. Thus, mass production eliminated product uniqueness.

This mass standardization is now reversing with additive manufacturing opening the door for mass customization - producing goods to meet individual customer's needs with near mass production efficiency. This is potentially closing the circle back to cottage industries where each item can be unique and tailored to the customer's needs. It also may decentralize manufacturing and, in the limit, allow making products in each home.

A prime example of this is Sally's experience. She lives in Santa Monica California and uses her bicycle for transportation. She wanted to decorate her bicycle (a mass produced product that looked like most other bicycles) with flowers. She wanted to be able to easily clip silk flowers to her handle bars. She was sure that there was a market for this concept that others would also want to individualize their bicycles with inexpensive, snap-on high quality silk flowers.

Figure 2: Another example of Pedal Petal silk flower

She was discussing this one morning over breakfast with her husband, Patrick. Patrick happens to be a design engineer with a 3-D printer in his home (part of his business). They decided to see if they could design, manufacture, advertise, sell and deliver clip-on silk flowers by dinner time.

They itemized the tasks and divided up the effort:

1. **Find a source for silk flowers.** They are commonly available at craft shops, sufficient for first products. This tasked needed to be done by lunch time so the clip could be finalized. Sally was responsible for this. Later she worked to find a source for bulk, lower cost flowers.
2. **Design the clip.** The clip is the part to hold the flower and to snap onto the handlebar. It had to grip standard size handlebars tight enough to hold the flow in any desired position and be removable without damage to the clip, flower or bicycle. Patrick could rough this in before lunch and finalize an initial concept as soon as Sally completed Task 1.
3. **Make first prototype.** When Tasks 1 and 2 were sufficiently complete, a prototype could be printed. Patrick could do this after lunch.
4. **Iterate until clip worked** (as described in task 2) and looked good. Both Sally and Patrick are responsible for this.
5. **Find first customers.** Sally needed to set up a business on Etsy (www.etsy.com). Etsy is a marketplace where people connect to buy and sell unique goods. She also needed to develop a name for her product.
6. **Manufacture first products**.
7. **Take first order**.
8. **Ship by dinner time**.

An ambitious project plan.

The design and manufacture of the clip

Patrick needed to design a clip that could:

- Snap onto most standard handlebars (15mm to 30mm in diameter) repeatedly without deformation or breaking
- Be easily secured on the handlebars
- Be easily fastened to flowers
- Have first examples that could be made on his Systems Cube II 3Dprinter. The Cube is limited to ABS and PLA materials. Full production on other AM machine to be determined later.

Patrick's solid model of his design for the clip is shown in Fig. 3. The left bulb section is designed as a receptacle for mounting the Silk flower stems. He designed it so a hot-melt glue gun could be used for final fixing of the flower to the clip. The rolled features on the left side make attaching easy. To lock the clip to the handlebars he decided to use a simple rubber band. One rolled side is nearly closed to hold it as shown in Fig. 4, with the other side serving as a hook. Fig. 4 is a photograph of the final product with ribbon covering the glue joint.

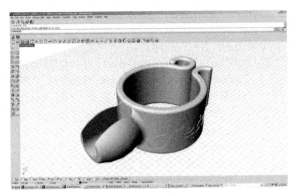

Figure 3: Solid Model of the clip

Patrick was limited in materials as the Cube only uses ABS and PLA. So he designed the clip with no stress concentrations under tension to minimize the opportunity to fracture with repeated use and chose ABS, the tougher of the two materials.

Manufacturing of the first eight examples were done on his home printer by mid afternoon. This gave him and Sally sufficient examples to test and one to ship to the first customer in time for dinner.

The Following Days

Larger batches were later made on a sPro60 Selective Laser Sintering (SLS) system with Duraform EX (Nylon 11). Patrick does not have this machine in his home as it is a production system, but he was able to use the services of cubify.com to scale up production and get a large batch of product. Nylon parts made with this system can be annealed to further strengthen them. Annealing was also done at home by boiling in water as can be seen on Sally's stove top in Fig. 5.

Figure 4: Back view of the final product

Figure **5**: Annealing nylon parts

Moving beyond the original manufacturing run of Nylon parts, Patrick cost compared production on SLS system with injection molding and found that at a unit time of approximately 30 seconds per part this approached the injection molding rate but without the tooling lead time and the flexibility to easily change part features.

Using the additive manufacturing process Sally can easily change features on her Pedal Petals, adding more features and testing them with customers.

Pedal Petal is truly a twenty-first century cottage industry.

Conclusions

Additive manufacturing let Sally and Patrick conceive, design, manufacture and ship a product between breakfast and dinner. The AM technology is clearly a game changer for the way products will be designed and delivered in the future.

Acknowledgements

Patrick and Sally-Anne Dunne of Pedal Petals assisted in writing this case study.

A Soft Ride at BikeE
A Case Study for *The Mechanical Design Process*

Introduction

During the 1990s, BikeE was one of the top manufacturers of recumbent bicycles in the world. A recumbent bicycle is one where the rider is seated or lying down. In 1995 BikeE introduced the AT model with an active rear suspension. Later that year the AT was named one of Bicycling Magazine's Best New Products. From 1996 - 2002, BikeE sold over 15,000 ATs. Many are still on the road today.

This case study was used as an example in the 3rd edition of *The Mechanical Design Process*[2]. What makes it a unique case study is that it takes a product from need, through concept development to detail design. It uses many best practices detailed in the book.

The Problem: Design a rear suspension for a recumbent bicycle

The Method: The engineers used the best practices in *The Mechanical Design Process* in great detail.

Advantages: The use on many design best practices (e.g. QFD, concept development, component development and DFA) led to an award winning bicycle suspension design.

Figure 1: The BikeE AT

---

[2] The author of this book was also the lead engineer for the development of the BikeE AT.

Background

The BikeE Corporation manufactured bicycles that allowed the rider to pedal in a seated position. This position is more comfortable than that with a traditional bicycle, with little pressure on the neck, wrists, and arms. On these bikes, it is easy to see the scenery or traffic as the rider's head is upright. Additionally, since the frontal area is small, they can be very fast. These types of bicycles are easy to maneuver and fun to ride.

The BikeE CT was the first model produced. It was introduced in 1992. The CT (Fig. 2) is characterized by a cantilevered rear stay (the part between the aluminum extrusion body and rear wheel). This stay is cantilevered to provide a little flexibility much like a diving board, a simplistic suspension system. It was important in the design of this bike to have as much flexibility as possible in the rear stay as 75 percent of the rider's weight is over the rear wheel. Additionally, some limited shock absorption is offered by the flexibility of the wheel and the foam cushion on which the rider is sitting.

Figure 2: The BikeE CT

In 1995 BikeE undertook a project to design an actual suspension system for the rear wheel. The resulting product was the AT. This case study describes the evolution of the AT.

The Design Plan

BikeE was a small company developing its first bicycle rear suspension system. The design team consisted of a design engineer, a product manager, a technician, manufacturing manager, machinist, materials specialist and an industrial designer. The core of this team was the design engineer, the product manager and the manufacturing manager. The core team drafted a list of tasks, as shown here. Note that these are fairly generic.

1. Generate engineering specifications.
2. Design two concepts.
3. Develop P1 prototypes.

4. Test P1 prototypes.
5. Select one concept.
6. Develop P2 prototypes.
7. Field test P2 prototypes.
8. Generate product documentation.
9. Produce production plan.

These, they put together in a Gantt Chart, Fig. 3 to show the relative timing for the tasks.

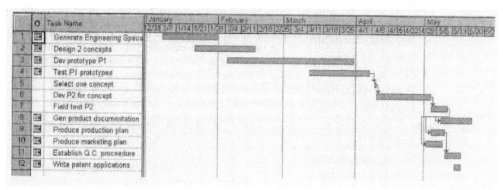

Figure 3: Gantt Chart for BikeE AT Tasks

QFD Development

To ensure that they understood the problem, the team developed a Quality Function Deployment (QFD) chart (Fig. 4).

For the BikeE suspension system, the main customers were bicycle riders; generally both the purchasers and the users of the product. In interviewing current riders of BikeE products, the team realized that there were two types of riders for the new product. First was the rider who rides solely on streets. The BikeE was initially designed as a road or commuting bicycle. However, there were some riders who wanted to go off-roading, so the second group of rider-customers were those who wanted to ride on rough roads or trails.

An additional group of customers considered were bicycle shop sales people and mechanics (often the same people). These people had to be enthusiastic to sell the product, answer questions about it, and repair it. Within the company, manufacturing, assembly and shipping personnel were also considered as customers but not shown in the example QFD.

To gather customer information for the BikeE suspension system, the team developed a survey and distributed it to current BikeE owners. Below is a sample of the questions included in the survey:

Q1. How many miles do you ride your BikeE each week? (Circle the best choice.)
1. <5 miles
2. 5-10 miles
3. 10-30 miles
4. >30 miles

Q2. What surfaces do you often ride on? (Circle all that apply.)
1. Smooth pavement
2. Rough pavement
3. Gravel
4. Packed dirt paths
5. Forest trails

Q3. If your BikeE had a rear suspension system, what types of surfaces would you ride on? (Circle all that apply.)
1. Smooth pavement
2. Rough pavement
3. Gravel
4. Packed dirt paths
5. Forest trails

Q4. If your BikeE had a rear suspension system, how often would you expect to adjust or maintain it? Consider the maintenance similar to checking and adding air to your tires. (Circle the most frequent acceptable time period.)
1. Never
2. Once every 3 months
3. Once a month
4. Once a week
5. Every ride

Q5. What is most important to you? (Rank 1 through 5.)
___ Smoothing road surface bumps (e.g., rough surfaces, manhole covers)
___ Absorbing pothole shocks
___ Low maintenance
___ Cool looks
___ Maintenance ease

Q6. What is your weight? ____

Q7. Describe your experience using your BikeE to go to work or school.

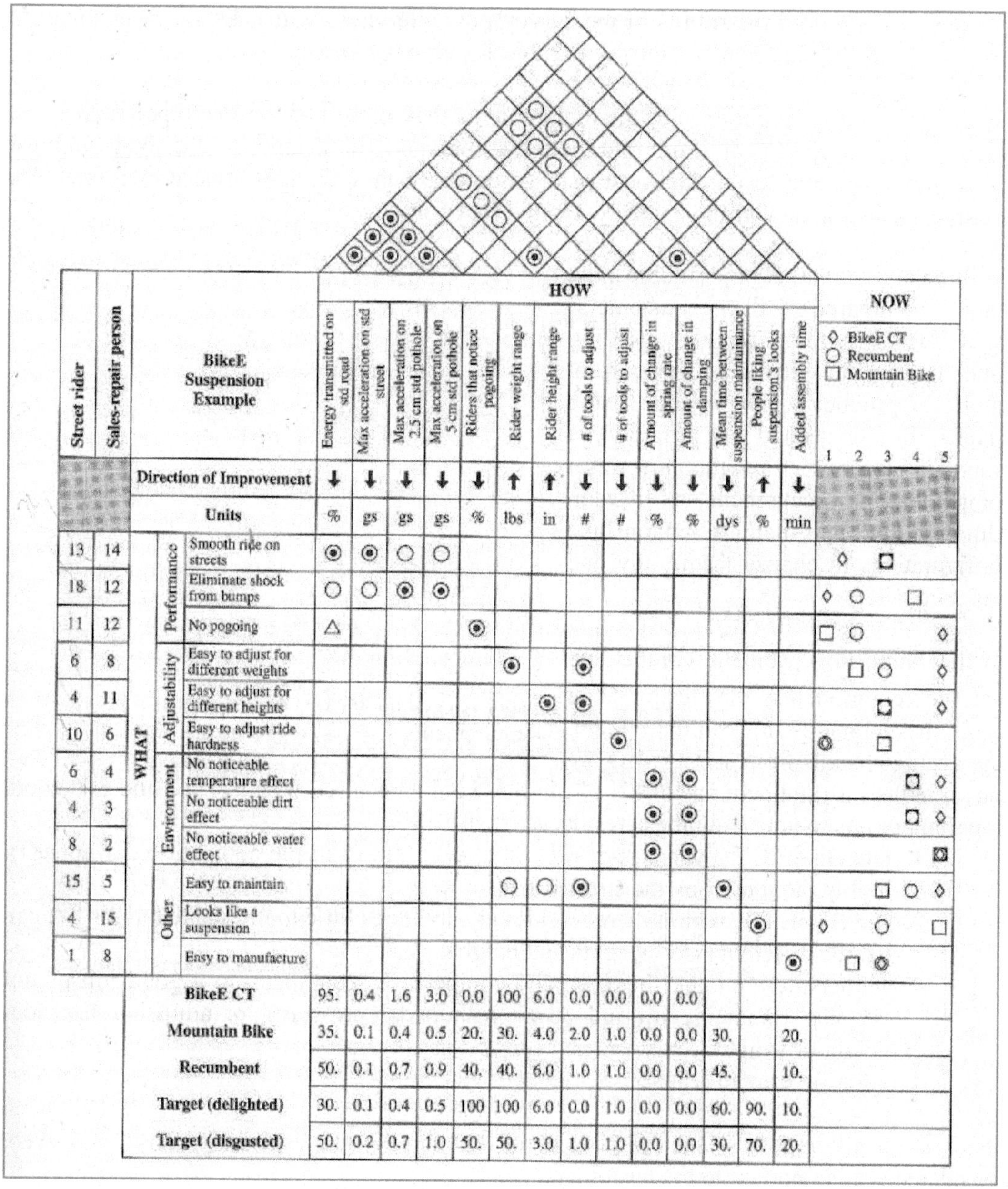

Figure 4: QFD for BikeE AT

BikeE Suspension Example

Direction of Improvement and Units for each HOW:

HOW	Direction	Units
Energy transmitted on std road	↓	%
Max acceleration on std street	↓	gs
Max acceleration on 2.5 cm std pothole	↓	gs
Max acceleration on 5 cm std pothole	↓	gs
Riders that notice pogoing	↓	%
Rider weight range	↑	lbs
Rider height range	↑	in
# of tools to adjust	↓	#
# of tools to adjust	↓	#
Amount of change in spring rate	↓	%
Amount of change in damping	↓	%
Mean time between suspension maintainance	↓	dys
People liking suspension's looks	↑	%
Added assembly time	↓	min

WHAT (Street rider / Sales-repair person):

Street rider	Sales-repair person	Category	Requirement	Energy transmitted on std road	Max acceleration on std street	Max acceleration on 2.5 cm std pothole	Max acceleration on 5 cm std pothole	Riders that notice pogoing	Rider weight range	Rider height range	# of tools to adjust	# of tools to adjust	Amount of change in spring rate	Amount of change in damping	Mean time between suspension maintainance	People liking suspension's looks	Added assembly time
13	14	Performance	Smooth ride on streets	◉	◉	○	○										
18	12	Performance	Eliminate shock from bumps	○	○	◉	◉										
11	12	Performance	No pogoing	△				◉									
6	8	Adjustability	Easy to adjust for different weights						◉		◉						
4	11	Adjustability	Easy to adjust for different heights							◉	◉						
10	6	Adjustability	Easy to adjust ride hardness								◉						
6	4	Environment	No noticeable temperature effect										◉	◉			
4	3	Environment	No noticeable dirt effect										◉	◉			
8	2	Environment	No noticeable water effect										◉	◉			
15	5	Other	Easy to maintain						○	○	◉				◉		
4	15	Other	Looks like a suspension													◉	
1	8	Other	Easy to manufacture														◉

NOW (ratings 1–5): ◊ BikeE CT, ○ Recumbent, □ Mountain Bike

Technical targets:

	Energy transmitted on std road	Max accel on std street	Max accel on 2.5 cm pothole	Max accel on 5 cm pothole	Riders notice pogoing	Rider weight range	Rider height range	# tools to adjust	# tools to adjust	Change in spring rate	Change in damping	Mean time between maint.	People liking looks	Added assembly time
BikeE CT	95.	0.4	1.6	3.0	0.0	100	6.0	0.0	0.0	0.0	0.0			
Mountain Bike	35.	0.1	0.4	0.5	20.	30.	4.0	2.0	1.0	0.0	0.0	30.		20.
Recumbent	50.	0.1	0.7	0.9	40.	40.	6.0	1.0	1.0	0.0	0.0	45.		10.
Target (delighted)	30.	0.1	0.4	0.5	100	100	6.0	0.0	1.0	0.0	0.0	60.	90.	10.
Target (disgusted)	50.	0.2	0.7	1.0	50.	50.	3.0	1.0	1.0	0.0	0.0	30.	70.	20.

The design team used the results of the survey plus interviews with bike shop personnel to develop the list of requirements shown in Table 1 and in the QFD.

To find the relative importance of the requirements they presented the list to potential street riders and sales-repair people and told them that they had 100 pennies to distribute amongst the requirements and to put the most pennies on those they thought most important. The average results for each group are shown in the first columns of Fig. 4.

To find the true market opportunities, the team compared the customers' requirements to three existing benchmarks. The first was the current cantilever product shown in Fig. 2 (the BikeE CT). The second was a traditional mountain bike system and the third was a competitor's recumbent system. Although there are many mountain bike configurations to choose from, only one was used here.

Smooth ride on streets
Eliminate shocks from bumps
Easy to adjust suspension system for different weight riders
Easy to adjust suspension system for different height riders
Easy to adjust suspension system for ride hardness
Easy to maintain
Looks like a suspension
Not noticeably affected by temperature
Not noticeably affected by dirt
Not noticeably affected by water
No pogoing[1]
Easy to assemble
Cost less than $50 to manufacture over the rigid rear fork
Weigh less than 400 grams over the rigid rear fork
Does not change bike height

[1] A bike that pogos moves up each time a pedal is pushed so it bounces up and down twice for each pedal revolution. Pogoing results from a poor design where the chain tension interacts with the suspension.

Table 1: Preliminary List of Customers' Requirements

To determine how well the competitors met the requirements, the design team used questionnaires to evaluate them. The average results from street riders are shown in the far right columns of Fig. 4. In these columns a score of 1 is bad and 5 is good. Important points to note from these results are that:

1. The BikeE CT gives a poor ride on streets, but even the competition is not very highly thought of by the street riders.
2. The BikeE CT, with its semi-rigid rear stay, does little to eliminate bumps but can handle different rider weights and heights.
3 Neither the mountain bike nor the competitive recumbent do a good job of not pogoing (bouncing up and down with pedal pressure), or adjusting for rider height or weight.
4. None are easy to adjust.

These were all important factors considered in the design, especially those that were considered very important to the customers.

The team then developed engineering specifications for the BikeE suspension as shown in the central column headings in Fig.4.

Some important points about these are:

1. In order to measure the "energy transmitted on a standard road," a standard road needed to be defined. Also, a method of measuring the energy content of the road and of the energy that was transmitted needed to be devised.
2. Some specifications are subjective. For example, it may be possible to actually measure "pogoing" and "looks", but it may be easier to use a test panel and note percentage of subjects who notice pogoing or like the looks.
3. The "# of tools to adjust" is listed twice. The first time, for weight and height, will have a target of 0. Height and weight could have been listed as two different measures, but they were combined, as the target is the same. The second, for adjusting the ride hardness, will have 1 as a target number of tools.

Conceptual Design

The design team took a careful, functional approach to the design of the swing arm. During the design of the semi-rigid rear stay for the CT model they had missed some functions and wanted to take more care there.

For the suspension system, the "most important" function can be worded in a couple of different ways. The team used "transfer forces between wheel, chain, and frame and absorb peak loads between wheel and frame," which is really two overall functions - transfer and absorb. These helped them define the boundaries of the system: the wheel, the chain, and the frame of the bike, as shown in Fig. 5; and that the primary type of flow is energy.

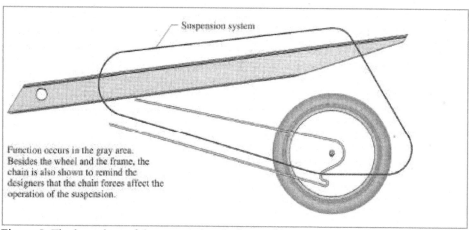

Figure 5: The boundary of the suspension design problem.

To help understand the function of the system, the design team drew a simple free body diagram (Fig. 6) based on Fig. 5. The arrows in Fig. 6 represent the forces due to the chain tension, the wheel pushing up, and the frame loading due to the rider's and frame's weight on the suspension system. This problem is essentially a two-dimensional problem as side loads are small on a bicycle. Note that during the design of the BikeE CT the force due to chain tension was not considered properly. It was later learned that the highest fatigue stress in the rear stay (i.e., the part that

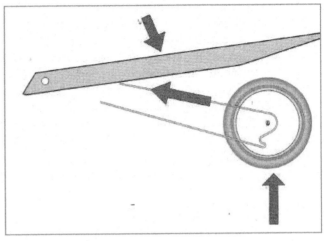

Figure 6: The forces on the suspension system

connects the frame to the stay in Fig. 2) was due to chain forces and not the vertical forces due the rider's weight and resulting vertical wheel force.

Based on this understanding, the team decomposed the main functions into sub-functions, as shown in Fig. 7. This diagram focused on energy transfer functions. The team decomposed the force from the wheel into as many functions as they could think of. Note that they focused on interfaces, how the energy gets in and out of the systems, and what has to happen internally to the energy. They divided the transfer of energy into concern for large bumps and small bumps based on their experience that it is difficult to get a smooth ride over little bumps and still have a system that can take the large hits. They also divided the energy storage from the energy dissipation, the spring from the damper.

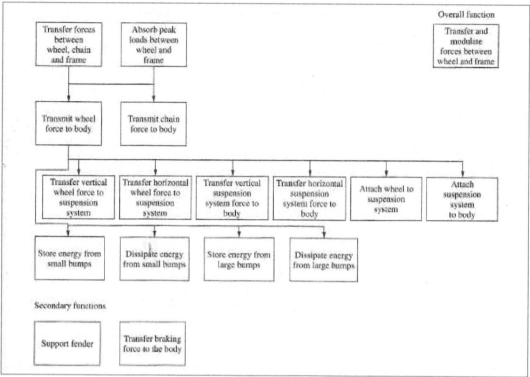

Figure 7: Functional decomposition for the suspension system

While developing the sub-functions they also remembered that the system to be designed would probably have to carry a fender and the brakes. They noted these as secondary functions.

The team used the functional model as a basis for a morphology (Fig. 8). Here, the ideas generated for each main function are shown.

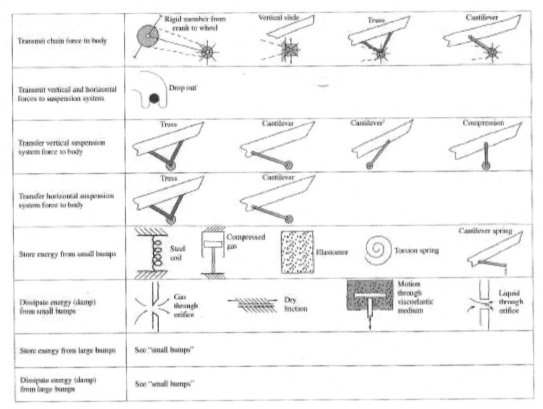

Figure 8: Morphology for the suspension system

Concept Evaluation

The BikeE team used a number of different methods to reduce the total number of concepts generated with the morphology. As they reviewed the results of the morphology they eliminated many of the potential alternatives because they couldn't pass a simple go/no-go screening. Thus, they combined their generation and evaluation efforts.

From the morphology, members of the design team developed conceptual sketches in their design notebooks (examples in Fig. 9). Each concept sketch was followed by notes abstracted from notebooks. These were broken down into two main sub-problems: the bicycle's structure and the energy storage/dissipation method. The sketches labeled with an "S" are structural and those labeled with an "E" are energy management. During the exercise of developing these concepts the team found that they learned much about the project:

- Three of the four structural concepts used pivots. The development of pivots requires a new sub-project complete with requirements, function consideration, and concept generation.
- The energy management with springs and dampers can be implemented with an off-the-shelf unit or made from basic components.
- The four structural ideas can be combined with the three energy management ideas in up to twelve different ways. They needed to reduce these to two ideas before prototyping.

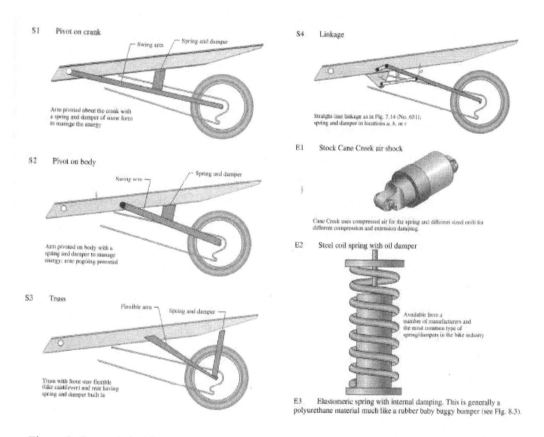

Figure 9: Concept sketches

They did take care not to eliminate too much, too soon, however. After using their judgment, they had four structures and three energy storage/dissipation methods. They thought that all 12 combinations of these were feasible-either conditional or worth considering.

In considering the twelve remaining concepts, the team compared each to customer's requirements in Fig. 4. None of the team had ever designed and built a suspension system before, so the technology assessment of each of the concepts was based on what they had read about. They concluded:

1. Probably any of the 12 alternatives could be made to meet most of the performance requirements itemized in the QFD, within what they knew at this point in the project.
2. The truss structure was eliminated as no one on the team was convinced that they knew much about the critical parameters, sensitivities, and failure modes of a flexible member on a bicycle.

Thus, this go/no-go screening reduced the number of concepts to use the air shock, coil/oil shock, or elastomer on either the pivot on the crank or a pivot on the body structures. In this case study we will focus only on the team's effort to decide on an energy storage/dissipation system.

The criteria for evaluation were developed from the customers' requirements in Fig. 4. The resulting list of eleven items is shown in Table 2. Some of the customers' requirements were left off or combined to keep the list to the most important for consideration.

Since the list was new, the team reevaluated the importance of the 11 criteria by allocating 100 points among them. The results shown in Table 2 are compromise values for the entire team.

Table 2: Decision Matrix for energy management system

	Criteria	Importance	Alternatives		
			Air shock	Coil/oil	Elastomer
1	Smooth ride on streets	13	+		−
2	Eliminate shocks from bumps	14	S		−
3	No pogoing	7	S	D	−
4	Easy to adjust for different weights	11	+	a	−
5	Easy to adjust ride hardness	6	S	t	−
6	Good environmental insensitivity	8	S	u	−
7	Easy to maintain	9	S	m	+
8	Looks like a suspension	5	S		S
9	Easy to manufacture	14	S		+
10	Understand critical parameters	8	S		S
11	Latitude and sensitivity known	5	S		S
		Total +	2	−	2
		Total −	0	−	6
		Overall total	2	−	−4
		Weighted total	24	−	−36

The three alternatives considered for energy management were an air shock, a coil/oil system, and an elastomer. Evaluation was done with the Coil/oil as datum. The other two options were rated using the -, S, + scale. The air shock had the best scores, implying it was the best concept.

Note that in order to evaluate the elastomer, a test bike was built (Fig. 10). This bike not only allowed testing the elastomer, it also allowed the first experiments with the pivot on the body. Even though it was early in the project to build a bicycle, it was necessary because knowledge about the elastomeric suspension was just too low to evaluate without it.

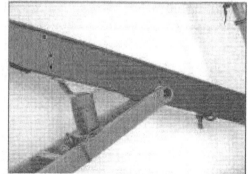
Figure 10: Elastomer test bike

Product Design

Most of the effort to design this system was accomplished by a single engineer. Only a small part of the work will be presented, as the entire history of the design process is too long. This discussion is organized, namely, from constraints to configuration to connections to components. Integrated into this flow is concern for materials and manufacturing processes, as these were considered these concurrently. The first step taken was to understand the constraints on the system.

Functional constraints were developed above, so initial concern here was with the spatial constraints. As shown in the layout drawing (Fig. 11), the other components that constrain the suspension system are the *frame*, the *wheel*, and the *chain*. The wheel is shown in two different positions, fully extended and fully compressed. The chain is represented as an envelope determined by what gear the bike is in and the compression of the suspension system. Although not shown, the suspension system could not interfere with the *seat*, the *ground,* and the *rider;* and the suspension system had to support *a fender*. All the items that are shown in italics provide spatial constraints on the suspension system. Initial focus was on the constraints shown in the layout drawing and then they were checked as the product evolved to ensure it cleared what it had to and that the accessories could be mounted.

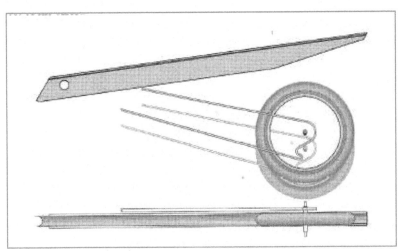

Figure 11: Layout drawing

It was quickly noted that there were only two major components, the swing arm (the member that connects the frame, the air shock and the wheel) and the air shock itself. However, by the time this product made it to production there were over 15 components on the BOM (Bill of Materials). In other words, to make this a producible system, each major component needed many pieces. A key decision while laying out the product was where to position the swing arm and air shock relative to each other and on the frame.

Many configurations were tried (one is shown in Fig. 12), and for each the forces, stresses, the loads on the air shock and the amount of potential pogoing due to the chain not being aligned with the swing arm was studied. There was a tradeoff between the performance of the system and its ability to fit within the envelope defined by the other components. This was especially challenging in considering the chain, as the swing arm needs to clear it regardless of what gear the bike is in, how far the suspension is deflected and the chain deflection when going over a bump.

Fig. 12 shows the result with the swing arm roughed in and the positions of it and the air shock figured out. The swing arm used in this drawing is similar to that used on the experimental bike (Fig. 10). The layout drawing in Fig. 12 helped ensure that the configuration met all the constraints.

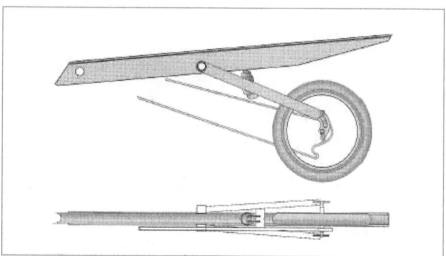

Figure 12: One layout of the swing arm

The next challenge was to design the connections or interfaces between components. This proved to take the greatest part of the effort, which is common. Essentially there are four connections, each will be discussed and the unique problems faced with each will be emphasized.

- *Connection between wheel and swing arm.* This was the easiest joint to design as all common bicycle wheels have the same axle diameter and connect to bicycles with a component called a "dropout" (if the nuts that hold the wheel on are loosened, the wheel drops out of the bicycle frame). The geometry and shape of a dropout was well refined and there was no reason to change it.

- *Connection between the swing arm and the air shock.* This too was well refined in that the air shock manufacturer supplied bushings to allow the shock to pivot relative to the swing arm (one degree of freedom). The drawing in Fig. 13 shows the bushings that allowed the shock to pivot. The only design issue was how to get the load from the swing arm to the air shock, a problem addressed shortly.

- *Connection between the frame and the air shock.* This also makes use of the same one-degree-of-freedom interface on the air shock, as shown in Fig. 13. The big challenge here was in ensuring that the forces transmitted through the shock could be distributed in the aluminum wall of the frame, a topic considered later.

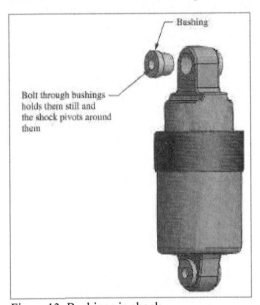

Figure 13: Bushings in shock

- *Connection between the swing arm and the frame.* This was the most time consuming interface designed during the project. It was essentially a one degree-of-freedom pivot but it had to connect the two blades of the swing arm to the thin aluminum wall of the frame. There are many ways to make a single-degree-of-freedom joint. Those that were considered included mounting the pivots to the side wall of the frame, the bottom of the frame and in a structure outside the frame. Additionally, the actual pivot bearings considered were a ball bearing (making use of an off-the-shelf assembly used for mounting the cranks and pedals), a bronze bushing, a plastic bushing, and a flexible joint (a solid piece of material that flexes within its elastic limit to provide a pivot that works over a small angle). The actual placement and configuration of the connection is dependent on the pivot types.

Next, the components were developed - the material between the connections. There are too many to detail all of them here and so only a few are discussed.

The Swing arm, a Component

The swing arm design was based on the cantilever structure from the CT model that worked fairly well and looked good. Additionally, the prototype, shown in Fig. 10, using a simple straight member seemed to work well, but this needed extensive analysis and testing. Thus, even though most bicycles are based on a truss and trusses are strong structural shapes, this effort focused on a cantilever swing arm.

For structural members in bending, the strongest shape is an I-beam. For the swing arm, this is not practical. The rectangular section used in the prototype (Fig. 10) and earlier models approximates an I-beam in that much of its material is far from the neutral axis, but the industrial designer on the team felt that this looked crude and not sleek enough for this new product. She wanted an oval shape. Shapes that were considered are shown in Fig. 14, note that the clearance of the chain forced special attention to the interference between components.

Other sub issues included:
- Verification of the strength of the swing arm.
- Developing a manufacturing method to make oval tubes out of round ones or finding a vendor who could supply them at a competitive price. A decision had to made whether to make the tubes or buy them.
- Developing a manufacturing method to blend the oval shaped swing arms with the flat dropouts, the fittings that mount to the wheel.

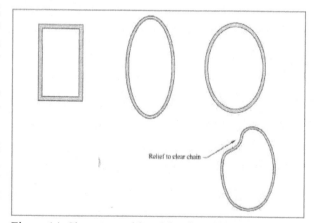

Relief to clear chain

Figure 14: Shapes considered for the swing arm

The load in the swing arm needed to be transferred to the air shock. This was a difficult issue - designing a simple component in a compact space, that cleared the air shock and chain, transferred the loads between the swing arm and the shock, and was easy to manufacture. Two ideas were explored (Fig. 15): the "spider" and the "tube." Each of these introduced its own unique challenges in connecting the swing arm to the air shock.

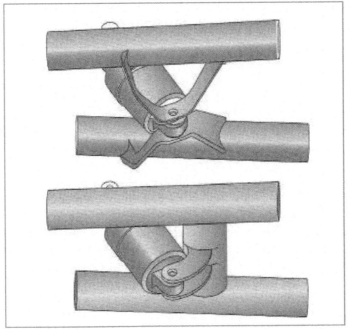

Figure 15: Connections for the air shock to the swing arm. The "spider" on top and the "tube" on the bottom

The strength and ease of manufacturing these configurations were the main requirements. The important points he addressed were:

- The spider configuration is essentially two opposing trusses, thus most of the structure is in tension. The shape of the truss or spider legs is not straight, as they have to clear the chain. The depth of the legs is large to distribute the stress over a broad section of the swing arm to keep the stresses down.
- The tube configuration makes the stresses much more

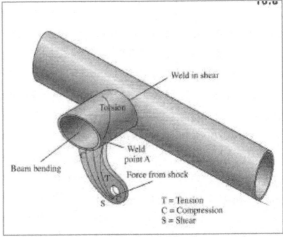

Figure 16: Force flow in the air shock connection

complex than the spider. The force flow is as shown in Fig. 16. The air shock puts the finger it is attached to in bending. This makes the weld at point A very critical because the stress is highest at this point. The fingers put the tube in both torsion and bending. Thus, the tube was sized to take these loads and the welds designed.

The team analyzed and tested each of the components developed here to find the modes of failure. In every case they wanted be sure that any failure would result in a safe situation. For example, even though the swing arm was designed to take the high forces generated from heavy people riding off curbs and other similar expected uses, they considered what would fail if someone rode the bike off a cliff (which someone later did).

Integral to their thinking were the materials to use for the components. Because BikeE and other bicycle manufacturers traditionally used 4130 steel for frames, they chose to use this material. However, the member of the design team that represented manufacturing wanted to consider brazing rather than welding the parts together to ease the manufacturing time required. They spent significant time studying how the heat used in each of these processes would affect the strength in high-stress areas.

Product Evaluation

Three of the Engineering Specifications on the QFD (Fig. 4) were for vertical accelerations during different riding conditions: maximum acceleration on a standard street, maximum acceleration on a 2.5 cm standard pothole, and maximum acceleration on a 5 cm standard pothole. Translating these specifications to a P-diagram, the street surface or pothole is the input signal, the maximum acceleration is the quality measure and the targets are, as given in Fig. 17.

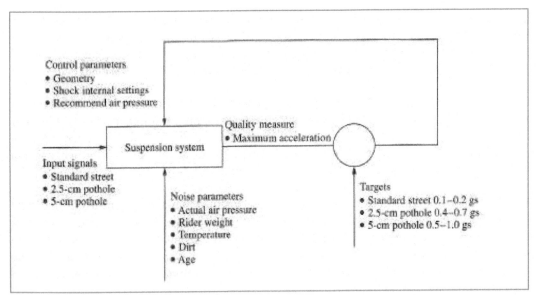

Figure 17: P-diagram for BikeE suspension

Also shown in the P-diagram are the control and noise parameters. The design team had control over the dimensions of the swing arm (e.g., its length, the location of the air shock on it), some of the internal settings in the air shock and the recommended air pressure for the shock. What they did not have control over was:

- The actual air pressure in the shock
- The weight of the rider
- The temperature
- The dirt buildup on the shock
- The age of the shock

These parameters are all noises. The designers knew that the customer would consider the system a quality product if it met the engineering specifications and was insensitive to these noise factors.

They realized that the first two "noises" can be somewhat managed through the geometry of the shock and the swing arm, but they also had to put limits on rider weight and suggested pressures in the owners' manual. The manufacturer of the chosen shock had done a good job making the unit temperature, dirt and age insensitive. BikeE also planned to sell shock rebuild kits as an aftermarket item[3].

---

[3] A Crane Creek shock on the author's bike lasted 15 years with heavy use before it needed a rebuild.

The engineers at BikeE had some simulation capability, but this was only sufficient to ensure that the performance was in the range of the targets. They felt that the best results could be found with physical hardware. Thus they built a test bike and instrumented it for measuring acceleration. They also set up a test track with 2.5- and 5-cm potholes. Tests were performed with riders of differing weights and with pressures different than those recommended. They also experimented with dirt on the shock and with heating and chilling it. Their goal was to find the best configuration of the parameters they could control and be insensitive to the noises.

Cost

The cost to make and sell a BikeE AT is broken down in Table 3. As can be seen, purchased parts are over half of the direct cost. To make all the custom parts and assemble those with the purchased parts took about 9 hours. BikeE's margin, the amount they made on each bicycle sold to a dealer, was 29% or $171.

Table 3: Cost breakdown for a BikeE AT.

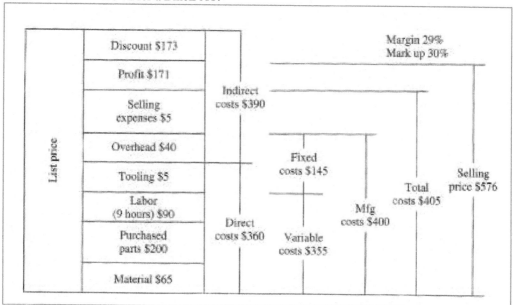

Design for Assembly

BikeE engineers knew that one way to improve the margin was to make custom components easier to assemble. One of the most complex assemblies on the early Bikes was the seat frame, Fig. 18. This frame had nine different components requiring 20 separate operations to put together. These included positioning and welding operations, which took 30 minutes. They knew that there was room for improvement, thus they undertook a Design for Assembly exercise as part of the AT development.

The new seat, Figure 19 has only four components requiring 8 operations and about eight minutes to assemble. The team used a template to compare the old and new seats. The form for the new seat is shown in Fig 20. The total score is 86/100 where the old seat was 70/100. The main difference between the two was in the Overall Assembly measures. For the old seat, the part count was poor and there was no base part for fixturing. Further, part mating was improved.

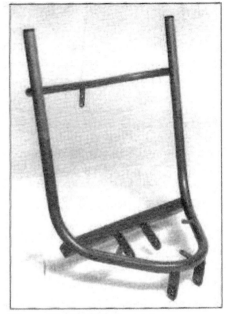

Figure 18: Old BikeE seat frame

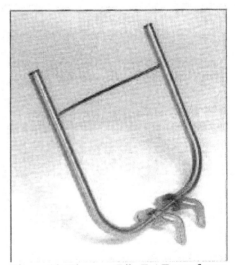

Figure 19: The new BikeE AT seat frame

(DFA) Design for Assembly		
Individual Assembly Evaluation for: New BikeE Seat		Organization Name : BikeE

	OVERALL ASSEMBLY		
1	Overall part count minimized	Very good	6
2	Minimum use of separate fasteners	Very good	6
3	Base part with fixturing features (Locating surfaces and holes)	Outstanding	8
4	Repositioning required during assembly sequence	>=2 Positions	4
5	Assembly sequence efficiency	Very good	6
	PART RETRIEVAL		
6	Characteristics that complicate handling (tangling, nesting, flexibility) have been avoided	All parts	8
7	Parts have been designer for a specific feed approach (bulk, strip, magazine)	Few parts	2
	PART HANDLING		
8	Parts with end-to-end symmetry	All parts	8
9	Parts with symmetry about the axis of insertion	All parts	8
10	Where symmetry is not possible, parts are clearly asymmetric	All parts	8
	PART MATING		
11	Straight line motions of assembly	All parts	8
12	Chamfers and features that facilitate insertion and self-alignment	Most parts	6
13	Maximum part accessibility	All parts	8
Note: Only for comparison of alternate designs of same assembly		TOTAL SCORE	86

Team member: Bob	Team member:	Prepared by:	Date:
Team member:	Team member:	Checked by:	Approved by:

The Mechanical Design Process
Copyright 2013, David G. Ullman

Designed by Professor David G. Ullman
Form # 21.0

Figure 20: DFA Template for the new seat

Conclusions

The BikeE team developed the AT model using many best design practices in *The Mechanical Design Process*. This resulted in an award winning and successful product.

Links

- US Patent 5,509,678; Recumbent Bicycle
- DFA Template link see book web site.

CPSIA information can be obtained
at www.ICGtesting.com
Printed in the USA
BVOW07s1456280218
08729BV00004B/30/P